To John and Anne-Marie,

...ion.

teve.

The Present of Things Future

The Present of Things Future

EXPLORATIONS OF TIME IN HUMAN EXPERIENCE

Thomas J. Cottle

Stephen L. Klineberg

THE FREE PRESS
A Division of Macmillan Publishing Co., Inc.
NEW YORK

COLLIER MACMILLAN PUBLISHERS
LONDON

The Free Press
A Division of Macmillan Publishing Co., Inc.
866 Third Avenue, New York, N.Y. 10022

Collier-Macmillan Canada Ltd., Toronto, Ontario

Library of Congress Catalog Card Number: 73-5292

Printed in the United States of America

printing number
1   2   3   4   5   6   7   8   9   10

Library of Congress Cataloging in Publication Data

Cottle, Thomas J   1937–
    The present of things future.

    Bibliography: p.
    1. Time. I. Klineberg, Stephen L., 1940–
II. Title. [DNLM: 1. Behavior. 2. Time. BF468
C849p   1974]
BF468.C6   153.7'53   73-5292
ISBN   0-02-906820-7

For
Claudia and Jason Cottle
and
Geoffrey and Katharine Klineberg

Contents

Preface

Time as experience and as dimension has long held a central place in the speculations of philosophers and in the interpretations of theoretical physicists. Only recently, however, have scientists who study behavior begun to focus explicitly on this fundamental aspect of human experience.

Implicitly, temporal experiences always have been present, always have provided the underlying structure and logic of human behavior. Time interposes itself into the thoughts and dreams of every one of us, calling as it does for the reconciliation of stability and change, of being and becoming, of whether to seize the present moment and live it fully or to postpone the gratifications of today in the interest of long-range goals and distant plans. The experience of time, as behavioral scientists too have recently begun to affirm, fashions the process of living and defines the inevitability of death. Indeed, no student of human behavior can afford to ignore the manner in which men and women perceive the dimension of time.

This book explores the psychological, sociological, and cultural forces that shape such temporal perspectives. Since we cannot begin to cover in a book of this size all aspects of social science inquiry into these perspectives, our concern will focus on one central dimension of the experience of time. Life evolves toward the future, and all human beings, individually as well as collectively through the cultural symbols and social institutions of their society, develop means and styles of managing the uncertain and the unknown. In these pages, we seek to learn above all how it is that people come to anticipate what the future holds for them, to define the major forces that appear to shape the contours of these anticipations and that underlie their power to influence present moods and actions.

In the behavioral sciences, the significant writings on these issues are widely dispersed in isolated fragments of a still uncompleted picture, more often characterized by elegant speculation than informed by adequate research. There is much, however, that has been learned. In these chapters, we shall try to bring this scattered work together into a preliminary formulation of the nature of temporal experience, with particular reference to the causes and consequences of the images that individuals form of their personal futures.

Within the context of the published research we have presented our own studies of time perspectives. Our choice of methods of inquiry is deliberately eclectic. We have sought in particular to wed two distinct approaches—each with its own strengths and limitations, each helping to fill the gaps left by the other. The vast majority of the research studies cited throughout this volume represent the "scientific" approach—the approach designed for experimentation and replicability, for accurately measuring rates and patterns, for testing relationships among variables and for reporting their levels of statistical significance. The data that emerge from the systematic sampling of populations, from the use of careful controls and structured questionnaires, are clearly indispensable if the science of human behavior is to advance. Without these techniques, we would have no way of generalizing with any assurance from the rich uniqueness of the individual case to the power of the general principle.

If, however, in an overly narrow devotion to the scientific method we were to limit ourselves to such data alone, we would risk losing sight of the human experience that underlies our conceptual and statistical abstractions. Thus we have undertaken a second form of research called the Life Study. Here, the rigor and replicability of the structured questionnaire is replaced by the openness and unpredictability of conversation and involvement, as four insightful individuals speak of their own encounters with time.

Building on these two approaches and developing against the background of the existing research, our work unfolds in three major sections. Part 1 establishes the foundation for the more focused inquiries of later chapters. Two central questions guide this preliminary inquiry. First, what are the central factors that appear to shape the images that individuals form of their personal futures? Second, under what conditions do individuals come to trust those images as guides for their behavior? Chapter 1 explores evolutionary processes and chimpanzee behavior, the functions and malfunctions of the human brain, the structure of attitudes and the nature of anxiety. For a different kind of preliminary perspective, we turn in the second chapter to the words of a seventy-nine-year-old man who, in the last weeks of his life, tells of his own personal overview of time.

Part 2 explores the maturational or developmental aspects of temporal experience. In particular, Chapter 3 examines the impact of cognitive growth and social learning as childhood yields to adolescence in Western society and the personal future acquires a new sense of reality. Chapter 4 deals with the acquisition of sex-role identities. Here a study is presented of the distinctive time perspectives inherent in the social roles through which young men and women come to define themselves. The life study in this section (Chapter 5) is of a nineteen-year-old Boston woman. Focusing on her childhood and adolescence and what they mean for the future, her words increase our understanding of these developmental influences on temporal perceptions.

Part 3 examines the cultural and social structural contexts through which each of us constructs his own experience of time and imagines what his personal future will be like. Chapter 6 explores the time perspectives that appear to be characteristic of "traditional" societies and examines the human "obstacles" that are thought to impede progress toward modernization in developing countries and toward the eradication of poverty in American society. Nowhere, perhaps, are the processes that link the individual with the social more clearly seen than under conditions of rapid change, when new conceptions of the personal future assume major importance. The research presented in Chapter 7 explores some psychological consequences of social change in Tunisia, and in particular its effects on the time perspectives of adolescents. The final life study, in Chapter 8, reveals the personal expression that such socio-cultural forces assume in the lives of a middle-aged couple living in a working-class area of Boston.

The three life studies and Chapter 4 were written by Thomas J. Cottle; the remaining four chapters were contributed by Stephen L. Klineberg. It was out of our appreciation and respect for each other's contrasting approaches to similar phenomena that this book was conceived. It is our hope in these pages to demonstrate how our respective methods of research, each possessing its own technical limitations, its own spirit and sense of ethics, may be beneficially joined in exploring the nature of time in human experience.

Acknowledgments

We would like to thank The Field Foundation, and in particular Leslie Dunbar; many residents of the Great Lakes Naval Training Station; the Laboratory of Social Psychology of the University of Chicago; the Department of Social Relations of Harvard University; the Education Research Center and the Medical Department of the Massachusetts Institute of Technology; the Alliance Française; the Center of International Studies at Princeton University; the American Philosophical Society; the Social Science Research Council; and the Office of Family Planning in the Ministry of Health of the Tunisian Government.

In addition, we wish to express our gratitude to Nelson Aldrich, Jr., Robert J. Beck, William Bezdek, John M. Butler, Betty Clinton, Anthony Costonis, Chandler Davidson, Carl N. Edwards, Jacob W. Getzels, Gerald Goldberg, Lawrence Grauman, Jr., Charles P. Henderson, Jr., Peter Howard, Otto and Selma Klineberg, Donald N. Levine, Joseph Lynn, John F. Marsh, David C. McClelland, Stanley Milgram, Kenneth Monts, William Morgan, John Muller, Irving Naiburg, Gerald M. Platt, Joseph Pleck, Taoufik Rabah, Devra Rosenberg, George Rowland, Jack Sawyer, Mady Shumofsky, Alvin H. Strelnik, Fred L. Strodtbeck, Robert Wallace, and of course, Kay M. Cottle and Margaret K. Klineberg.

PART
1

Preliminary Perspectives

1
The Future
in the Present—Its Creation
and Its Power

> There are three times: a present of
> things past, a present of things present,
> and a present of things future.
>
> St. Augustine (*Confessions*)

Time permeates all of human experience. When an infant is born, the joy he brings to the present is informed by the promise of his future. As a child grows older, he inevitably learns to attune his behavior to the rhythms of physical time—the alternation of day and night, the recurrent cycles of hunger and satiation, activity and rest—and he gradually becomes aware as well of his own participation in the irreversible progressions of biological time—the experience of maturation and the knowledge of inevitable death. Gradually, too, he develops a conception of himself within the framework of his social world, and of his life as a whole in both its lived and its as yet unlived portions. As he encounters the conventions of his social world, he learns to give particular meaning to these universal experiences of time. Above all, he comes to create increasingly realistic and powerful images of what his future may hold, and he trusts those images as guides for his behavior.

In the pages that follow, we shall try to uncover the central processes through which a person anticipates his future and the forces that shape his anticipations and influence their power to control his behavior. This first chapter explores the published research in the behavioral sciences in an effort to develop a preliminary formulation of the psychological forces that underlie the causes and consequences of the images people form of their personal futures. Our discussion begins with an inquiry into what makes it possible in the first place for a person's anticipations of the future to enter into his cognitive world, there to shape his experience of the present.

THE MENTAL IMAGE

Anticipating the personal future implies at least two central processes. One must first of all be able to imagine events that have no concrete reality, and one must then confer a sense of reality on them by attributing them to the realm of the possible as distinguished from the purely fanciful world of wish-fulfillment. The forces that shape these processes are extremely complex. Indeed, they will occupy our attention throughout most of this book.

The sheer capacity for mental imagery represents one of the most important culminations of evolutionary development. Our nearest animal relatives show surprising imaginative and anticipatory capabilities. There is much, however, that escapes the bounds of their capacities, and an examination of primate behavior may help to define the contours of what is unique about man's experience of time and the future.

Imagery in Chimpanzees

There is continuity in phylogenetic development. Among its most striking dimensions is behavioral rigidity: The farther one descends the animal scale, from man through the primates to lower mammals and beyond, the more rigid and fully predictable is the animal's response to the stimuli it encounters in its environment. A squirrel finding its first nut on a bathroom floor, for example, immediately runs through its instinctive burying program: It scratches at the tiles as though it were digging, bears down on the nut with its nose as if to push it into the floor, and completes the sequence with covering movements—all in seeming indifference to the fact that the nut remains as fully exposed as before (see Hess, 1962)[1]. It is as if the squirrel's brain in this instance were operating as a telephone switchboard whose sole function is to connect incoming sensory messages to preprogrammed motor responses. The animal appears on the scene with a complex and coordinated series of fixed actions, the nut acts as a "releaser," and the burying program unfolds in a blindly automatic sequence at the first sight of an appropriate nut.

Fixed action patterns of this sort become increasingly rare as one ascends the animal scale. In higher organisms, communication appears to go on within the brain, as well as into and out of it (Hebb, 1966). This phenomenon has to do with the development of "association areas" in the brain, of neural cells that are not given over to receiving incoming stimuli or coordinating outgoing responses, but are available instead for registering the learning of complex behaviors and for representing the goals to be achieved through that activity. In higher animals, then, behavior derives from mediating processes in the brain itself rather than as a direct reaction to immediate stimulation. A period of delay now intervenes between stimulus and response, and the eventual behavior is no longer predictable simply from the nature of the stimulus input and knowledge of the physiological state of the organism. It is this progressive increase in the proportion of "association areas" in the brain that underlies the striking evolutionary shift from primarily reflexive to primarily purposeful behavior.

[1] Our method of citation in this volume is the common one of providing the author's last name and date of publication in the text itself. The List of References at the end of the book supplies the complete bibliographic information.

Examples of such "mediated" behavior abound in published studies of chimpanzees. The capacity for delayed response, for remembering the location of a reward after a single experience and without visual cues, is perhaps the clearest indication of "imagery" in animals. Yerkes (1943) demonstrated how accomplished chimpanzees are in this ability. An ape that has once observed a banana being placed in one of several unfamiliar boxes and is then taken out of the room can remember the exact location of the banana and will go to the correct container immediately on returning to the room, if the delay interval does not extend beyond a minute or two. Moreover, if during the period of delay the experimenter substitutes a piece of lettuce for the banana, the ape will usually refuse to take it. Shrieking in anger, the animal typically searches through the other containers and around the room, looking for the lost banana. We know from other studies that were the animal not at that moment expecting to find its favorite fruit, it would have taken the lettuce without hesitation. It is difficult to explain why it refuses to do so now, unless we assume that the ape was somehow able to keep in its mind, throughout the delay period, the specific image derived from having previously seen the prized banana placed in one of the boxes.

The chimpanzee's capacity for mental imagery, however, shows some surprising limitations. In one of his experiments, Yerkes put the banana into one of two differently colored boxes while the ape was watching. He then switched the position of the containers while the animal was out of the room. When brought back, the chimpanzee went directly to the place where it saw the food concealed and persisted in searching for it there, as if only position could be represented in its mind, and color or shape were entirely forgotten. Yerkes (1943, pp. 178–179) commented as follows on this surprising discovery:

> To us it seems almost incredible that with both boxes before it, but interchanged in position, the animal should go where the food had been concealed, in spite of the altered appearance of the box, and there search persistently for its expected reward. Under like circumstances a person would notice at once the

changed appearance of the food box and look about for the
original box. . . . We finally were forced to admit that our
subjects either failed to perceive the essential clue, or were
unable to hold it in mind because they lacked a symbol or
representative process comparable with our word "green."

The absence of language thus severely limits the range of experience that chimpanzees are able to represent in their association cortex.
Their capacities are nonetheless sufficiently developed to permit some
remarkable manifestations of anticipatory and purposeful behavior.
Hebb and Thompson (1968) described the intriguing example of
captive chimpanzees that, after seeing a visitor arrive on the scene,
will often slip quietly over to the faucet, fill their mouths, and return
to the front of the cage, there to wait serenely until the visitor is
within range, at which point he is suddenly drenched with water. Even
without the symbolic potentialities of human language, chimpanzees
are clearly able to anticipate a goal and to plan a series of complex
actions before actually carrying them out.

We now know that the capacity for purposeful behavior, for actions guided solely by images of future events, is not a purely human
attribute. It remains true, however, that only in humans do we encounter anticipatory and purposeful behaviors involving periods not
just of a quarter of an hour but of weeks or years or even decades,
and we need to search further for the bases of this unique advantage.

The Capacity for Temporal Integration

The processes of remembering and anticipating are possible only
when mediating mechanisms in the brain enable the organism to
imagine objects or events that are not a part of its immediate sensory
environment. The chimpanzee, we have learned, is indeed capable of
carrying away from situations of the kind created in the Yerkes laboratory what Mowrer and Ullman (1945) characterized as "some inner
change or state which 'stands for' the response which it will later make
when it re-encounters the same situation." From the behavior de-

scribed by Hebb and Thompson, we learn that the ape is also capable of visualizing an event that has not yet occurred, but there are clear limitations in both aspects of its symbolic activity. Not only is the chimpanzee unable to represent mentally certain key features of its environment, such as color or shape, but it also seems unable to anticipate future events unless the relevant external stimuli are physically present. We have no reason to suppose that those captive chimpanzees were thinking beforehand that it might be fun to squirt water on visitors. The anticipations that subsequently guided their behavior emerged only after they had perceived the relevant external stimulus —the unsuspecting visitor.

Arieti (1947) suggested that behavior of this sort should be clearly distinguished from true "anticipation," by which he meant "the capacity to foresee or predict future events even when there are no external stimuli which are directly or indirectly related to those events." It is this capacity that appears to develop only in human beings. It may be that chimpanzees are so limited in the length of time over which they appear able to plan their actions partly because distant anticipations are rarely connected directly with present perceptual events.

In his classic study on *The Mentality of Apes*, Köhler (1925) suggested that one of the chief differences between anthropoids and "even the most primitive human beings" is to be found in the extremely narrow span of "the time in which the chimpanzee lives." He concluded as follows:

> The lack of an invaluable technical aid (speech) and a great limitation of those very important components of thought, so-called "images," would thus constitute the causes that prevent the chimpanzee from attaining even the smallest beginnings of cultural development (p. 238).

All other animals, even the brightest apes, live primarily in the present. In contrast, it is extremely difficult for an adult human being to immerse himself completely in present experiences for any prolonged period. The associated events from both the past and future continually intrude into consciousness, and there they are integrated into the contemporaneous patterns of action and meaning. It is this process of "temporal integration" (Hearnshaw, 1956) that causes the stimuli of perceptual experience to lose their predominance. Only in humans do we find that *representations* of objects and events play as great a role in determining behavior as does their direct perception.

Human experience is predominantly symbolic and representational. It is above all *construed* experience, and a large part of its meaning derives, as we shall see, from its temporal connectedness. Furthermore, as Murray (1959) argued, much of human behavior can be explained only "by reference to persistent 'self-stimulation' in accordance with a plan of action." Underlying this distinctively human reality is a person's prodigious capacity for manipulating symbols and creating images, for transforming the physical contours of present experience into symbolic conceptions, and for linking them with images of the past and the future. A person's capacity to shape and give meaning to his life and his world far transcends the few mental representations available to the brightest chimpanzees.

THE CREATION OF THE FUTURE

If anticipations in human beings rarely derive from present perceptions alone, how then are they created? Our first clue comes from Murray's concept of "self-stimulation" and his insistence that much of human behavior is motivated by "plans" that are inwardly generated rather than externally evoked. As a person acquires linguistic forms, he learns to turn his language back upon himself, as it were, to construct an image of himself and of his place in the social world. Self-awareness implies a knowledge not only of where a person is now, but also of where he was in the past and where he expects to be in the future (Hallowell, 1955). As a person gradually assimilates his changing recollections into a sense of identity and of self-continuity, anticipations of the future become an inextricable part of the meaning he confers on present experiences. If this is so, then it must be at least partly out of our reconstructions of the past that we are led to create our images of the future. As Kelly (1958) pointed out, "man exists primarily in the dimension of time," and he "makes of himself a bridge between past and future in a manner that is unique among creatures."

The Past in the Future

In a study of a small village in Normandy, French anthropologists Bernot and Blancard (1953) discovered that the peasants in the village planned their lives in terms of a far longer perspective on the future than did factory workers recruited from other communities. The authors suggest that for the workers, events in the relatively distant personal past had no continuity with their present circumstances; as a result, the future they envisioned also appeared as an unconnected image of a better life rather than as a goal toward which they could direct their efforts. They showed little awareness of the intervening steps that might some day make that image a reality. The peasants, on the other hand, had a far deeper sense of continuity with the past and with the land that had been their source of livelihood for generations. They viewed their future as a further extension of that same continuity. Having been provided for by his father and grandfather before him, the peasant was also likely to plan in terms of providing for his children when their turn came. Humans do indeed appear to make of themselves a bridge between past and future: The deeper their ties with the past, the longer their perspective on the future.

More direct and quantitative evidence for this relationship can be found in many studies in the behavioral sciences. When people are asked to look at a series of ambiguous pictures and to tell a story about each (a technique known as the Thematic Apperception Test, or TAT), the temporal span covered by the action they describe suggests the way they deal with events in time. If the situation depicted on the TAT card represents "the present," then all story content that refers to a period prior to the situation on the card might be scored for "retrospective time span," and that part of the story that deals with events occurring after the "present" moment might indicate "prospective time span." The association between these two scores has been found repeatedly to be of high statistical significance (see Epley and Ricks, 1963; Wohlford, 1966, 1967). Adolescents and adults who show a greater concern for past events in their relation to the present are also likely to trace the consequences of present behavior into the more distant future.

Similarly, Graves (1961) asked 133 high school students to list ten different things they anticipated in the future and ten events that they had experienced in the past. After both listings, they were asked to indicate how long they thought it would be before each future event came about and how long it had been since each past event had occurred. Again, a high correlation emerged between the general remoteness of the future events anticipated and of the past events remembered. It seems, as Jacques (1956, p. 91) suggested, that "the length of time into the future that likely consequences can be traced is related to the capacity mentally to organize previous experience."

This capacity will reflect not only individual differences in personal styles of temporal experience, but also situational factors and the speed of social change in general. As the past grows increasingly remote and discontinuous with the present, the future, too, is likely to be conceived as unpredictable, its images unsafe as guides for current actions and meanings. The relationship between past and future also runs in the opposite direction, for recollections are open to continual reinterpretation and reassessment. In the world of common sense, the past is often assumed to be fixed and immutable as against the ever-changing flux of the present and the supple fantasies of the future. As Peter Berger (1963, p. 57) argued, however, common sense appears to be quite wrong in this assumption: "At least within our own consciousness, the past is malleable and flexible, constantly changing as our recollection reinterprets and re-explains what has happened."

The Future in the Past

As a person alters his conception of his future direction, as he assumes new social roles, for example, and revises his sense of self, he is likely to reconstruct his past as well. "What an individual seeks to *become*," Rollo May (1958) suggested, "determines what he remembers of his *has been*." Thus it is that personal as well as national histories continually call for reassessment and revision in the light of changing conceptions of the present. And just as each new generation of historians feels compelled to rewrite a nation's past, so each significant change in a person's life impels him to re-examine his past experiences and their implications for his future. "The novelty of every future," G. H. Mead (1932) wrote, "demands a novel past." As a person bridges the past and the future via the emergence of a sense of personal identity, these two temporal "worlds" interpenetrate and shape each other.

There is another sense in which this interpenetration seems to occur. John Cohen (1964) pointed out that "the further ahead one looks, the more the vision of a millennium resembles the golden age of the mythical past." This paradoxical phenomenon is beautifully illustrated in Robert J. Lifton's (1964) analysis of the various ideologies which Japanese youth have fashioned in their efforts to manage the vast historical discontinuities created in the aftermath of the Second World War. The "transformationist" is wholly committed to radical change in the pursuit of distant goals, but underlying and somehow fused with his dominating valuation of the future, Lifton discovered, is a deep nostalgia: He "embraces a vision of the future intimately related to, if not indeed a part of, his longing for a return to the golden age of the past."

In summary, it is primarily in a person's efforts to comprehend his present experiences that his conceptions of both past and future are shaped. If, in his retrospections, he can find a sense of continuity and of orderly, predictable change which accounts for his current circumstances, he is likely to project into the future a sense of new possibilities that derives from these interpretations of past and present experiences. Images of the personal future are thus created in the process of "temporal integration," in the connections that people themselves impose on their experience of time. This process, however, does not in itself account for the power that images of the future possess to motivate behavior. "More than an ability to remember," Tomkins (1962, p. 131) wrote, "is involved in the development of anticipation. What is remembered must also be compelling here and now. The individual must care if he is to act on his anticipations."

THE IMPORTANCE OF PRESENT EMOTIONS

If the words a person says to himself in imagining his future are to motivate behavior, they must somehow generate in themselves appropriate emotional experiences. In our search for insight into the underlying dynamics of this process, we shall return briefly to the fields of evolutionary theory and physiological research.

A Possible Organic Basis

As we have pointed out, one of the key features of phylogenetic development is the progressive increase in the relative size of brain areas that are not given over to registering incoming stimuli and coordinating outgoing responses. In the specific evolutionary development that led from our ape-like ancestors to modern man, parts of that association cortex located in the anterior portions of the brain, primarily in the area known as the frontal lobes, increased in size more markedly than others. Recent evidence suggests that this region is particularly involved in the elaboration of imagery and planning. Damage to the frontal lobes appears to disrupt the operation of the mediating processes that underlie the capacity of higher animals to imagine events that do not actually exist.

Chimpanzees, we know, can easily sustain a delay of at least ninety seconds after having seen some food placed under one of two inverted cups and then covered by a screen. They can maintain a faithful image of the situation they have seen, and respond correctly once the period of delay is over. But when Jacobsen (1936) removed their frontal lobes, he discovered that an interval of just five seconds disrupts their capacity for delayed response. Several years later, however, Malmo (1942) demonstrated that if a lobotomized ape is kept in total darkness during the period of waiting, it will respond as accurately as it did before the operation. Images can be maintained, then, even without the frontal lobes; but these areas of the brain appear to perform the vital function of filtering out all the distracting stimuli that develop during the normal period of delay. The frontal lobes are somehow involved in maintaining the functional connections between memory and desire, between the emotion aroused by having seen the food and the image of the lever to be pushed or the cup to be removed in order to obtain the reward.

Research with human beings, although often inconsistent, appears to support this conception of frontal lobe functioning. The evidence suggests that our ability to develop and maintain orientations toward distant future events may depend not only on our capacity for symbolic behavior, but also on the functioning of the greatly enlarged frontal lobes that characterize the human brain.

In 1942, Freeman and Watts published the results of their controversial operations on chronic psychotic patients for whom all other treatment had failed. The patients underwent prefrontal lobotomies, an operation that involves severing the main connections between the frontal lobes and the thalamus, and they showed marked improvement: Obsessive and compulsive behavior was alleviated; chronic tensions and anxieties were relieved. In their place, however, new symptoms gradually developed. The post-operative patients were strikingly deficient in initiative and the capacity for planned action. They appeared to show little concern for the consequences of their behavior, and they had difficulty restraining themselves from responding impulsively to the immediate stimuli in their environment. The authors were reluctantly led to the conclusion that "lack of foresight seems to be one of the most characteristic manifestations of deficiency in frontal lobe activity" (p. 209).

Several later studies demonstrated in more quantitative fashion the extent to which prefrontal lobotomy focuses a patient's attention on the more or less immediate present. Jones (1949), for example, found a significant decrease after the operation in the proportion of patients' statements that dealt with "future-permanent" rather than "immediate-temporary" goals, and an increase in the percentage of stories they completed with ambiguous or missing endings, suggesting a disinclination to look to the future. Similarly, Petrie (1952) reported an intensive investigation of the effects of lobotomy in severely neurotic patients. She found that "the range of time with which the individual is concerned has shrunk, so to speak, and is now more or less confined to that which is here and now" (p. 111). After reviewing some of this research, J. Campbell (1954) concluded that "following prefrontal lobotomy, the patient's abstract concept of himself continuing in the future has been destroyed."

The evidence appears to be generally consistent in suggesting that the frontal lobes are implicated in the ability to resist immediate environmental stimulation, to maintain a response set over a ninety-second delay in the case of the chimpanzee, and to interpose a consideration of long-range past and future events in the case of the human being. The most plausible explanation seems to be that these areas of the brain somehow underlie the capacity for images of absent events to generate experiences of pleasure or discomfort, thereby motivating present behavior. Indeed, as Freeman and Watts (1942) themselves observed:

> It is especially the inner sensations, ideas, recollections, ambitions, and disappointments, the regrets for the past and the fears for the future, that are particularly affected by pre-frontal lobotomy. The capacity for imagination is still present, and certainly not sufficiently reduced to render the patient helpless, and affec-

tive responses are often quite lively, so that it is not a lack of
either one alone, but rather a separation of one from the other
where they concern the individual himself (pp. 303–304).

In all probability, it is this separation of image from emotion in
lobotomized patients that accounts for the striking reduction in their
tendency to anticipate the personal future. They "can often solve
complicated problems in the intellectual field where multiple choices
are offered, but their inability to foresee the effect upon themselves
in the relation to their environment is particularly outstanding" (*Ibid.*,
p. 313).

To summarize the results of our discussion thus far, it would
appear that a person's ability to act in the present in the light of his
anticipation of relatively distant future events entails three central
processes that are uniquely developed in human beings: (1) the prodigious capacity to manipulate symbolic representations of reality, and
thereby create images of absent events and believe in their validity;
(2) the ability to integrate these symbolic representations into ongoing
patterns of action and meaning, bringing images of the future into
an implied continuum with conceptions of past and present experiences; and (3) the present experiencing of pleasure or discomfort
generated solely by these symbolic representations. The evolution of
enlarged frontal lobes in the human brain appears to be a part of
the organic basis for this third essential process. The lack of foresight
pointed out in clinical descriptions of lobotomized patients strongly
suggests that unless images of the future give rise to present emotions,
anticipations will have little effect on a person's behavior. Our next
question, then, has to do with how mental images of this sort engender
such emotional experiences.

The Linking of Emotion and Image

Dollard and Miller (1950) provided one answer to this question
in the language of learning theory: Symbolic stimuli, like any others,
acquire "reinforcing" power through the process of "secondary con-

ditioning." If words, in the form of parental promises, for example, are regularly associated with "real" rewards, the words themselves will eventually acquire the capacity to reward performance and ensure a child's "good" behavior. In this way, the words that a person says to himself may come to be rewarding or punishing in themselves. Thus, "The disturbing thought that he may not be able to provide for his children's education may help to motivate a father to buy insurance and to keep working hard in the shop or office" (p. 107).

How have such thoughts in fact acquired their "disturbing" power? Is it really because the symbols through which they are expressed have become imbued with "secondary reinforcing power" by virtue of their previous association with real negative consequences for the individual? Has the father in the above example ever in fact been unable to provide for his children's education and suffered for it, with the result that he will now do whatever he can to avoid a recurrence of that experience in the future? Probably not. Much of human learning occurs vicariously through symbolic experience and empathic communication without the necessity for direct personal knowledge. It also occurs through the development of self-conceptions and through efforts to protect one's self-esteem. That father need not have actually experienced the unpleasant possibilities he now envisions; he need only know that others have suffered that particular fate and be capable of imagining how he would feel were he in their position. This suggests that anticipations of the future will motivate behavior to the extent that a person projects an image of himself into the situations he envisions, and in this way comes to experience *now* some of what he would experience if he actually found himself in those situations.

Most difficult decisions essentially involve a selection among alternative futures, and they are usually reached through this same process of self-projection. British economist G. L. S. Shackle (1958, pp. 41–42) described the decision-making process this way:

> The enjoyment or satisfaction which the decision-maker seeks
> to maximize by his choice of one action-scheme rather than

others is a pleasure of the imagination. . . . The decision-maker will be attracted towards that action-scheme which gives him the most enjoyment by imaginative anticipation of its possible success.

When upper- or middle-class adolescents are faced with choosing their life's vocation, some try to imagine themselves in various careers, and then select the career that provides them now with the greatest sense of personal satisfaction. Others tend more or less to drift into occupations without the kind of foresight that might ensure a better choice. The relationship between adequacy of vocational choice and the general tendency to anticipate relatively distant future events was tested by Sattler (1964). The respondents were twenty-eight counselor trainees who were given the Kuder Preference Record to measure the degree of similarity between their current interest patterns and those of successful practicing counselors. In addition, they were asked to list ten different events that they thought might happen to them in the future. The trainees who listed more distant anticipations also expressed patterns of interest significantly more like those of practicing counselors.

The Case of the Psychopathic Personality

What can happen to behavior when anticipations of the future are unaccompanied by present emotional experience is suggested in clinical descriptions of the "psychopathic personality." Problems of diagnosis have long plagued research efforts in this area, for the term has been used as a kind of wastebasket category to designate any long-standing maladjustment that fails to meet the more restrictive criteria of mental retardation, psychosis, or neurosis (White, 1964). Yet out of this process of exclusion a clinical picture emerges of one of the most fascinating and terrifying types of social deviance. The psychopath is a "rebel without a cause," a "moral imbecile" who seems able to commit any act with hardly a twinge of remorse, a loner who seems incapable of real affection for others or of lasting attachment to any code of values whatsoever.

When psychopaths turn to crime, they differ markedly from most criminals in their seeming inability to organize their efforts in the pursuit of any long-range goals. Their crimes appear to be wholly impulsive, with little evidence of a thinking process interposed between an idea and its execution. As the McCords (1964, p. 16) noted, the psychopath "is a man for whom the moment is a segment of time detached from all others. His actions are unplanned and guided by his whims." Lindner (1944, p. 3) put it this way: "With the psychopath, as with the infant, determined progress toward a goal—unless it is a selfish one capable of immediate realization by a sharply accented spurt of activity—, the dynamic binding together of actional strands, is lacking."

Given his deficient capacity to experience in the present emotions generated by his anticipations of the future, neither the specific fear of punishment nor the more generalized anxiety caused by a realization of the risks of crime is likely to deter his behavior. One of the "most basic characteristics" of the psychopath, Cleckley (1959) wrote, is his "relative freedom from anxiety and apprehension." Indeed, when psychopaths are compared with other prison inmates, they score significantly lower on measures of manifest anxiety (Lykken, 1957), and they show far less fear arousal as they anticipate being subjected to electric shock (Hare, 1965).

In a fascinating report, Redl and Wineman (1951) described their attempts to help a small group of psychopathic children. They soon discovered the central role that time perspectives play in determining the impact of education or therapy:

> These children had not developed much of a realistic concept of "themselves in the future" so that there was little to appeal to, one way or another. . . . This seemingly minor part of their pathology constituted one of our major technical hazards, as becomes clear when we remember that promise as well as threat,

punishment as well as reward, encouragement as well as criticism, have a rather healthy relationship to time as a *conditio sine qua non*, as a prerequisite without which they make no sense at all or are even doomed to make things worse (pp. 142–143).

Since so much of this symptomatology resembles the clinical descriptions of lobotomized patients we examined earlier, it is hardly surprising that brain damage is suspected to underlie many cases of psychopathic behavior. At this early stage of research, however, purely experiential factors cannot be ruled out of the equation (Bender, 1950; Bowlby, 1951). Indeed, Goldfarb (1945) effectively demonstrated the role of experiential forces in his comparison of children raised from birth in inadequate institutions with those who were quickly adopted into foster homes. However such questions of causality are ultimately resolved, enough is now known about the nature of psychopathy to suggest the dire consequences both for individuals and for their society of a deficient ability to anticipate future events in terms of their personal consequences. Underlying this inability is a disruption of the crucial connection between images of the personal future and present discomfort or pleasure.

EXPERIENCING THE FUTURE

If anticipation thus depends on the capacity of images of the future to engender appropriate emotions, we might expect that when emotional energies are totally absorbed in present experiences or deadened by excessive fatigue, the future will largely disappear from consciousness, and anticipatory images will lose their power to motivate present behavior. Indeed, Osgood (1962) suggested that when emotionally driven, our time span contracts to the present moment; and Maslow (1962) described the "peak experiences" of creative people as an intense absorption in the present, when a person is most free subjectively of both the past and the future, most "all there" in the experience itself.

If the future thus gains no entry into a present of intense absorption, feelings of severe dejection or discouragement will also constrain the capacity to bridge the past and future. Minkowski (1958) described such periods:

> Our synthetic view of time disintegrates and we live in a succession of similar days which follow one another with a boundless monotony and sadness. With most of us, however, these are only transient episodes. Life forces, our personal impetus, lift us and carry us over such a parade of miserable days toward a future which reopens its doors widely to us.

When, on the other hand, images of the future do become part of our experience, they may decisively influence our emotional state. Farber (1944) discovered that the degree to which a prisoner suffers in the present seems to be related to such variables as uncertainty about his release, the hope of getting a break, or feelings of injustice, but not to factors in the immediate day-to-day situation, such as satisfaction with his prison job. The crucial dimensions underlying the degree of present suffering appeared to be those "involving time perspective, particularly the future outlook."

The same investigator (Farber, 1953) found that preference for days of the week among college students was similarly affected by their perspective on the future, with Monday the least preferred day and Saturday the most preferred—even among students who had classes that morning. Moreover, Sunday was slightly more disliked than Friday, presumably because the pleasantness of that day of freedom was soured by the unpleasant reality of an impending week of classes. And May (1953, p. 262) emphasized the degree to which current experiences are enhanced if a person can connect his present actions with cherished future goals:

> Hope in its creative and healthy sense—whether it is hope for religious fulfillment or for a happy marriage or for achievement

in one's profession—can and should be an energizing attitude, the bringing of part of the joy about some future event into the present so that by anticipation, we are more alive and more able to act in the present.

What happens, however, if future images do not elicit joy? What happens if a person, in surveying his past and assessing his present, finds little reason to hope for better things to come?

The Phenomenon of "Antepression"

Bettelheim (1958) described the behavior of concentration camp inmates in Nazi Germany as they gradually came to adjust to the extreme agony and hopelessness of their situation: "The prisoners lived, like children, only in the immediate present; they lost the feeling for the sequence of time; they became unable to plan for the future or to give up immediate pleasure satisfactions to gain greater ones in the near future." Among American soldiers sent into battle against the German army in World War II, Stouffer and his colleagues (1959) observed a similar dynamic at work. A typical adjustment of the fighting troops, they wrote,

> involved a shortening of time perspective, living in the short run and focusing attention on means rather than ends. The hedonism of combat troops on leave is traditional. By banishing the long run to a status of limited reality, the combat man could to some extent escape facing the unpleasant probabilities and avoid a disorganizing level of anxiety (pp. 189–190).

Eson and Greenfield (1962) suggested that even in everyday life and among non-neurotic individuals "antepression" occurs at least as readily as does "repression." When their respondents, male and female, aged 10 to 65, were asked to list the things that they had recently thought or spoken about and to indicate their feelings about them, those items that referred to future events were particularly likely to be rated "pleasant." We rarely expect the worst from the future, and when we do, we tend to think about other things. Thus, as we shall see in the next chapter, even Marcus Simpson, knowing that his "days are numbered," still goes on making plans and imagining his future, a future that his mind fills not with progressive deterioration and senility, but with children.

Wohlford (1966) provided a rare experimental demonstration of this human tendency to suppress unpleasant prospects from one's anticipations of the future. College students were asked to describe in detail an unpleasant experience that they thought would probably happen to them in the future. Subsequently, when asked to list twenty topics that they had recently thought or spoken about, they mentioned significantly fewer future and more past events, compared with twenty other events listed before they were forced to envision an unpleasant future. When an individual confronts the probability of something unpleasant, he tends to stop thinking about the future, turning instead to the present or to the past, wherein events are less threatening if only because they have already occurred.

In summary, images of the personal future are such that they usually generate in themselves pleasant or unpleasant emotional states that intrude on current experience and motivate present behavior. It is precisely because of this link between emotion and image that some individuals are willing to postpone present gratifications for the sake of imagined future rewards. Others, however, tend to restrict their perspectives to the relatively immediate present in order to avoid confronting the prospect of an undesirable and seemingly unavoidable future. Disturbing thoughts of future prospects are motivating only as long as a person believes that he can do something now that might preclude their occurrence. A relatively restricted present orientation can operate as a "defensive" strategy, protecting a person from the pain induced by unpleasant anticipations. Images of that unhappy future may then be actively excluded from consciousness and relegated to a realm of pure subjectivity, safely unconnected with current realities. Under these circumstances, the span of future time that will be integrated with conceptions of the past and the present may be limited to those relatively short-range events that a person feels he can control. Alternatively, the time span may be restricted to those areas of life that offer more promising prospects or perhaps to images of far distant events, so far off that they remain safely detached from the disturbing realities of immediate experience. The dynamics of this important process are explored in the next section.

The Limited Span of Integrated Time

The process of temporal integration—linking images of the future with conceptions of the past and the present—may appear to disintegrate when future prospects are bleak, when the motivation to think about the future has therefore disappeared. Although personality differences are undoubtedly important in this connection, the social context in which a person finds himself may profoundly affect his images of time. The interpenetration of personal and social forces will be of central concern in Part 3 of this book. For now, it may suffice to indicate briefly the degree to which temporal integration is threatened by social instability, by external forces that render the future unpredictable and thus sever its psychological connection with the present.

Barker, Dembo, and Lewin (1943) discovered that children who were suddenly and arbitrarily prevented from playing with attractive toys regressed into far less constructive and imaginative play, as if they were no longer willing to organize their activities in terms of a more than momentary perspective. In a similar manner, when workers suddenly find themselves unemployed for a prolonged and indefinite period, they appear to lose a sense of the passage of time (Eisenberg and Lazarsfeld, 1938), and they tend to avoid most thoughts about the future, for anticipations only serve to renew their fears and anxieties (Zawadski and Lazarsfeld, 1935; Israeli, 1936). In summarizing the results of these and other studies, Lewin (1951, p. 114) wrote:

The unemployed man and even his children have been observed
to narrow their field of activities far more than economic neces-
sities require. Their time perspective seems to shrink so that
the behavior of the person is more dependent upon the immediate situation.

In Colombia, social instability has for many years been expressed
by outbreaks of banditry and violence. Lipman and Havens (1965)
interviewed sixty-five families who had personally been victimized
by "la Violencia" and had fled to the relative safety of Bogotá. The
attitudes and perspectives of these families were compared with those
expressed by fifty-nine other persons comparable in age, education,
income, and length of urban residence. Despite an average of ten years
in the relative calm of the capital city, the "Violencia" victims revealed
a far greater sense of insecurity and pessimism with regard to their
personal futures and a keener desire for life as it had been in the
relatively distant past. Having concluded that their social world was
ultimately unpredictable and threatening, they seemed to have little
inclination to look forward in time.

Under conditions of social instability, there may well occur what
Coser and Coser (1963) called "a mutilation or destruction of time
perspective." No longer able to maintain a meaningful sense of the
connection between his current situation and his images of the future,
a person makes little effort to bridge the past and the future in his
present experiences. By the same token, a relatively broad span of
temporal integration is facilitated by a stable social context that encourages the belief that present actions will have some measurable
impact on future outcomes, that what will happen in the future is
to some degree both predictable and controllable. As Finestone (1957)
observed, a personal orientation toward future events is the "subjective counterpart of a stable social order and of stable social institutions, which not only permit but sanction an orderly progression of
expectations with reference to others and to one's self."

Even in a stable and highly favorable environment, however, there
are clear limits to the span of future time that can be meaningfully

integrated into a temporal continuum. As a general rule, it seems safe
to suggest that the more temporally distant the future events envisioned, the more tenuous is their connection with the present, if only
because it becomes increasingly difficult to conceive realistically
events in the present that might dependably contribute to far distant
ends. Models of the decision-making process often recognize explicitly
that the decision-maker must impose a "temporal horizon" on his
deliberations, cutting off his vista of expectation at some definable
distance from the present. Beyond that point, consequences are increasingly difficult to envision realistically. Given enough time, almost
anything can happen, and calculating future events on the basis of
present trends becomes impossible (Roby, 1962).

Temporal integration fails if images of the future are restricted
to the relatively immediate consequences of present actions. It also
fails, paradoxically, when the future focus, in Frank's (1950) words,
"is so remote that it loses all potency over the present." In both cases,
the "real" future of weeks and months and years ahead has lost its
connection with the ongoing events of the present. The individual
is unable to bring the consequences of a contemplated action into
the psychological present, to compare and balance the anticipated
rewards and punishments before he begins to act (see Mowrer and
Ullman, 1945). To an outside observer, his behavior, attitudes, and
decisions may appear to lack a degree of reason or rationality.

The Temporal Setting of Attitudes

Back and Gergen (1963) found evidence in public opinion surveys
that people who felt they had passed their peak of mental ability and
those who expressed a sense of futility in their outlooks on the future
were far more likely to advocate immediate (or apocalyptic) solutions
to national and international problems, rejecting more moderate long-
range social programs. Similarly, Farber (1951) reported that college
students who said that their outlook on the next four years with regard
to their personal lives was "unsatisfactory" were significantly more

likely to favor an immediate showdown with Russia than were those with more optimistic perspectives on their personal futures.

A closed mind and a utopian ideology may characterize people who focus primarily on the far distant future. They have fashioned for themselves what Erikson (1959, p. 142) described as "an overly clear perspective on the future," one which suppresses the hard ambiguities and troublesome uncertainties that otherwise might engender intolerable anxiety. "The effectiveness of a doctrine," Hoffer (1951, p. 76) pointed out, "does not come from its meaning but from its certitude." A major characteristic of the "true believer" is his deprecation of the present through a glorification of the past or a preoccupation with the assured millennium to come in the distant future.

In a series of studies, Rokeach (1960) explored the relationship between closed-mindedness or "dogmatism" and temporal orientations. The espousal of a closed ideology, he suggested, is likely to be accompanied by a time perspective "in which the person over-emphasizes or fixates on the past, or the present, or the future without appreciating the continuity and the connections that exist among them" (p. 51). People who score high on his scale of "dogmatism" do indeed appear to be more preoccupied with future events than do low scorers (Rokeach and Bonier, 1960), and they evidence greater anxiety and a deeper sense of helplessness and despair in their outlooks on the future (Roberts and Herrmann, 1960).

This implicit relationship between anxiety and temporal defense calls for closer examination. More than any other human emotion, anxiety reflects the ambiguities and the complexity inherent in a person's efforts to come to grips with the dimension of time.

The Impact of Anxiety

A survey of the empirical findings concerned with the relationship between anxiety and the orientation toward future events suggests a picture of almost hopeless confusion. On the one hand, there is Siegman's (1961) discovery among Israeli undergraduates of a significant positive correlation between measures of manifest anxiety and the span of time covered by the action described in story-completions. This would seem to confirm his "simple assumption that the more one's psychological life space extends into the future the greater the uncertainty, and therefore, the greater the anxiety." We also have Rokeach's (1960, p. 367) suggestion, reversing the direction of causality, that "attempts to cope with anxiety should involve a de-emphasis on the present and a pre-occupation with the future."

On the other hand, there is Krauss' (1967) argument that highly anxious persons may try to lower their anxiety level by perceiving time relatively more in terms of the past and less in terms of the present and the future. Krauss and Ruiz (1967) tested twenty-three newly admitted psychiatric patients and found a significant negative relationship between manifest anxiety (measured by the same scale as in Siegman's study) and the use of present or future verb tenses in sentence-completion items: The higher the patient's anxiety level, the greater the proportion of past tenses and the greater his avoidance of present or future references. Epley and Ricks (1963), in an intensive study of seventeen college students, also reported a significant negative relationship between anxiety and the length of prospective time spans in TAT stories.

About fifty years ago, McDougall (1923) pointed a way out of this dilemma. He spoke of the "prospective emotions of desire" and suggested that among them anxiety occupies a central, deeply ambivalent position between the positive emotions of confidence and hope and the negative emotions of despondency and despair. A person in a state of anxiety is consumed by conflicting hopes and fears with regard to the future. As MacKinnon (1944) argued in elaborating this perspective, "The life space of the individual burdened with anxiety lacks a clear cognitive structure both in the psychological present and future. He doesn't know what to expect."

If there were none of this ambiguity, if expectations for the future were clear, we would expect either a restricted perspective as a protection against prospects that appear bleak and hopeless or an enhanced orientation toward future events when current experiences seem to justify optimism. The relationship between anxiety and time perspectives may be unusually complex precisely because anxiety represents neither despair nor confidence, but rather a deep sense of uncertainty. It is this ambivalence that is reflected in the conflicting findings just reviewed.

At the one extreme, it seems clear that the existence of anxiety itself presupposes an anticipation of the future, that its absence suggests a lack of concern with future events. As Maslow (1962, p. 14) noted, "Threat and anxiety point to the future (no future = no neurosis)"—a phenomenon we have encountered in clinical descriptions of lobotomized patients and psychopathic criminals. As long as the degree of anxiety remains moderate and a person still retains a feeling of at least partial control over his future, his sense of the connection between present and future may well be enhanced. Anxiety is likely to be a major motivation in a person's efforts to bring his anticipations of the future to bear on his present actions, to take steps in the present that may provide some reassurance regarding his prospects for the future. In the event, however, that that reassurance is not forthcoming, his sense of personal control will diminish, and his anxiety may become increasingly intolerable. His efforts will now be directed to ways of reducing that anxiety—by antepression, for example, or by fashioning an idyllic image of a distant future safely detached from the implications of present reality.

If we assume that Siegman's undergraduates ranged from low to moderate in their levels of anxiety and that the mental patients tested by Krauss and Ruiz were moderately to highly anxious, then the positive relationship between anxiety and future orientation reported in the first study and the negative relation found in the second no longer seem so surprising. Persons who are moderately anxious should indeed evidence a broader perspective on the future than those who experience very little anxiety or those who are burdened by intolerable degrees of that complex emotion.

In 1968, Ruiz and Krauss reported the results of their continued testing of all the new admissions to the psychiatric hospital in which they worked. There were now sixty-four subjects, and the greater range in levels of anxiety that they exhibited established a significant curvilinear relationship. Out of twenty-four sentence-completions, the sixteen patients scoring lowest in manifest anxiety completed an average of 4.5 items with verbs in the present or future, and the sixteen subjects scoring highest in anxiety completed 5.3 sentences in this manner. The two intermediate groups averaged 8.1 and 7.5 sentences invoking present or future actions.

The onset of anxiety clearly presupposes the anticipation of future events, and moderate anxiety may well facilitate the expansion of the scope of integrated time as a person searches for ways to control the future that he envisions. Excessive anxiety, however, may contribute to the dissolution of temporal integration. Images of the future may then be robbed of their connection with the present and hence of their power over present moods and actions. And a person may find himself confined to the present moment in a time that no longer "runs" (Binswanger, 1958).

There is another central consequence of the limited span of integrated time that we must now consider. When images of future events stand isolated and unconnected with present experience, imagination may be freed from the major set of forces that constrain it when dealing with realistic considerations. "The further ahead we look," John Cohen (1966) observed, "the more our vision is shaped by fantasy, just as a Proustian magic transforms the recollections of our distant past."

If "the future" refers to the span of time that is least determined and most open to new and real potentialities, it may also be the focus of wish-fulfilling fantasies that ignore the world of practical action and of human possibilities:

> We have to distinguish between pure or free imagination, unfettered fantasy, on the one hand, and on the other, that kind of constrained imagination which we call expectation. There are, of course, pleasures to be had from mere day-dreaming, but they are of a different sort from those of expectation. These

latter we may call enjoyment by *anticipation*. When we speak of what the decision-maker can visualize, we mean what he can *anticipate*, that is, what he can imagine without a sense of unrealism (Shackle, 1958, pp. 41–42).

The Constraints on Imagination

Without the ability to dream, to imagine events that have no concrete reality, there can be no creativity in general nor any possibility of creating the images of future goals that inform present experiences and give direction to current strivings. But unless the dream is tempered by reality, infused by logical forms of thinking and by the knowledge of real possibilities, it becomes a sterile escape from the time relationships of the real world. Eugen Bleuler (1912) drew an important distinction between two central modes of thought: *logical thinking*, "which corresponds to reality, is a reproduction in thought of the connections which reality offers"; whereas *autistic thinking* "is directed by strivings which disregard logic and reality":

> A lack of balance between these two extremes in an individual may produce either dreamers who only concoct combinations without regard for reality and do not act, or else sober realists who are so filled with realistic thinking that they live only for the moment, anticipating nothing.

As Lévy-Valensi (1965, p. 165) wrote, "La sagesse n'est pas de ne pas rêver, mais de féconder le réel par la puissance du rêve." [2]

Images of future events are likely to be tempered by realism to the extent that they are integrated with realistic conceptions of the past and the present, and as long as the events anticipated are felt to be controllable by present actions or predictable on the basis of

[2] In rough translation, "The wise course is not to abstain from dreaming, but to make the real world blossom through the power of the dream."

current trends. The "real" future, in other words, stops at some definable distance from the present moment, and anticipatory images are increasingly freed from the constraints of reality as they refer to events beyond that point.

But what is in this sense "distant" for one mind may be relatively "close" for another. William Stern (1938, p. 402) was one of the first psychologists to insist that objective measures of duration have little to do directly with the sense of the imminence of future events: "The future is *near* to the person," he wrote, "only when it is experienced in its personal significance for the *present*." The image of a "future" event may be viewed by one person as unconnected with the present and hence as free to vary with the whims of fantasy. Another may integrate that same event into his present experiences, construe it as a goal capable of realization, and define it in ways that reproduce what Bleuler referred to as "the connections which reality offers." If we accept Shackle's definition of "anticipations" as those events that a person can imagine without a sense of unrealism, we must also recognize that people will differ in their sense of what is "realistic," and thus in the events that may potentially be included among their anticipations of the future. And we must try to uncover the central processes that account for individual differences in this connection.

As is so often the case when one feels lost in the complexities of the human personality, an essential clue can be found somewhere in the work of Sigmund Freud. Writing in 1911, one year before Bleuler, Freud drew a strikingly similar distinction between two major types of mental functioning. He spoke of the "pleasure principle" through which the Id strives for immediate gratification, and contrasted it with the "reality principle," which describes the efforts of the Ego to delay impulsive gratifications until attainable goals are discovered in the real world of external constraints. When a person imagines the future, the contrast between these two modes of thought, according to Schafer (1958), "is one of megalomaniacal, unattainable, infantile conceptions vs. conceptions relatively more regulated by . . . realistic considerations of the possible rather than the exalted."

Freud suggested, moreover, that the emergence and eventual domination of the "reality principle" in mental functioning is a gradual development that proceeds through childhood and well into adolescence. It should follow that children, as they form their anticipations of the future, will be far less likely than adolescents or adults to maintain a clear and consistent distinction between unattainable wishes and realistic expectations. By examining the way conceptions of the personal future change as a child grows into adolescence, a great deal may be learned about the processes that underlie individual differences in the span of integrated future time and in the realism with which that future is envisioned. These are matters that we shall explore in Chapter 3.

THE EMERGING PERSPECTIVE

In this chapter, we have sought to provide a preliminary overview of research in the behavioral sciences as it relates to the causes and consequences of the images that men and women create of their personal futures. We noted the surprising capacities that chimpanzees possess for remembering past events and for anticipating the future. But the absence of language severely limits the range of experience that they can represent through imagery, and their anticipations of future events appear to depend upon the physical presence of external stimuli directly related to those events. It is man's prodigious capacity to create and manipulate symbolic representations of physical and social reality that underlies the uniquely human proclivity for anticipating distant future events and acting on them. Two fundamental processes emerged as centrally important in this regard. We have called them temporal integration and the linking of image and emotion.

Temporal integration. Although even the brightest apes live primarily in the present moment, it is extremely difficult for an adult human being to immerse himself completely in the present for any prolonged period of time. Associated events from the past and the

future continually intrude into consciousness, integrated into his contemporaneous patterns of action and meaning. In large part, this
process has to do with the emergence of self-awareness. As a person
gradually assimilates his changing recollections of his personal past
into a sense of identity and of self-continuity, anticipations of the
future become an inextricable part of the meaning of present experiences. The future is primarily created out of the implications of past
experience, of the sense of continuity and of orderly predictable
change that it may provide. Conversely, when altered circumstances
in the present create new conceptions of the future, reconstructions
of the past are likely to change accordingly.

Images of the personal future are thus created in the connections
that men and women impose on their experience of time. The process
of temporal integration does not in itself account, however, for the
motivating power that such images possess.

The linking of image and emotion. Feelings of pleasure
or discomfort generated solely by conceptions of future events occur
only as a person projects an image of himself into the situations he
envisions and in this way comes to experience in the present some
of what he would experience were those situations actually to occur.
The evolution of enlarged frontal lobes in the human brain appears
to be part of the organic basis for this essential process. Clinical
descriptions of lobotomized patients and of "psychopathic" criminals
emphasize the consequences for behavior of a future that produces
no emotion beyond indifference. Anticipations depend on the present
experiencing of pleasure or discomfort engendered solely by the images that the mind constructs.

If emotional energies are absorbed in present experiences or deadened by fatigue, the future largely disappears from consciousness.
Conversely, the feelings that are actually aroused by images of the
future may profoundly affect a person's present emotional state. If
a person despairs, however, and feels relatively powerless to control
his future, he is likely to make unconscious use of antepression.

In an effort to avoid the discomfort engendered by anticipating an unpropitious and seemingly unavoidable future, images of that future may be actively excluded from consciousness, and the span of integrated time may appear to disintegrate. A person may then restrict his perspective to those relatively short-range future events over which he feels he does have some control, or to those limited areas of life that offer more promising prospects, or perhaps to images of far distant future events, so far off that they remain safely detached from the present. As a result, the "real" future of predictable middle-range consequences may lose its psychological connection with the ongoing events of the present, and attitudes or decisions may appear to lack a degree of rationality. Temporal integration is also threatened by social instability, by external forces that render the future unpredictable and thus sever its subjective connection with the present.

Beyond the limits of integrated time, where images of "future" events stand isolated and detached from the actual experiences of the present, imagination is less likely to be tempered by realistic consid erations and may reflect instead the projection of wish-fulfillment. Future events that are psychologically "distant" for one person, however, may be "near" for another. People differ in the connections they perceive among events and in their sense of what is realistic. A great deal may be learned about the processes that underlie such individual differences as we explore in Chapter 3 how children's conceptions of the future change as they grow into adolescence.

This preliminary perspective on the forces that appear to shape our conceptions of the future serves as a foundation for the more focused inquiries in the chapters that follow. There is much of fundamental importance, however, that we have not as yet discussed. We have said nothing, for example, about the experience of aging, the awareness of death and the need to fashion a sense of "immortality," or the importance of retrospections for their own sake and for their own integrity as a life nears its completion. The next chapter is addressed to these issues.

2
Marcus
Nathaniel Simpson

"IF THE LORD HAS PATIENCE,
I'VE GOT A FUTURE"

The extensive research reviewed in Chapter 1 provides a comprehensive introduction to an examination of time in human experience. Perhaps the most fundamental notion we might take away from that chapter is that perceptions of the future necessarily connect with memories of prior events as well as with experiences of the present: "The associated events of past and future continually intrude into consciousness, there to be integrated into the contemporaneous patterns of action and meaning." Now, in the context of a life study, we wonder how this integration of the past, present and future in the passage of a human life might be explored. What connects the past to the future, and how is prior experience organized and reflected in anticipations and prospects? Can we trace these phenomena in personal expression? Can we discover in someone's words the symbolic representations of time periods and feel the poetry inherent in a concept that describes the flexibility of the remembered past?

In this chapter, a life study is presented of a man overviewing the time of his own life.* It is a special sort of overview, one that

*For a discussion of the methodology of life studies see the Appendix.

clearly links personal anticipation with present experience. It is an overview revealing a past and present being constantly revised as future inevitabilities draw nearer. Marcus Nathaniel Simpson was seventy-nine years old when we met. He was a patient in a Boston Veterans Administration Hospital living out, so he said, his last days. It is commonly believed that the aged experience no sense of the future, but live only with a sense of the past. In this absence of a future perspective, as we have just learned, they might be compared to psychopaths, lobotomized patients, and perhaps with inmates in concentration camps. But this conception is simply not true; it is merely another of the indignities we impose on older persons. For all human beings possess the capacity to foresee the future realistically, to know which of their wishes and expectations are likely to come true, and to understand how personal action shapes events and controls future outcomes. And as a result, to our last moments, the future has meaning and content.

The steps from the present to the future of Marcus Simpson's life were indeed short. Still, the oppressiveness of the hospital rooms in which we spoke never once destroyed his integrity nor the intellectual and emotional grasp of time that he had established. "Hope in its creative and healthy sense . . . ," Rollo May (1953) wrote, ". . . can and should be an energizing attitude. . . ." It was for Marcus Simpson. The intensity of the hope he countenanced was sufficiently strong to resist his incarceration and the continual appearance of death in his hospital ward. Everything one finds in those wards conspires to yield a sense of impending finitude, and only someone deeply committed to life could keep the imagery of death out of his construals of time.

There aren't too many men who could do what I can, look out across the entire span of my life, any life, that is, come to think of it, and see that precise way it all comes together. No sir, you won't have heard these things before. Makes me sad, too, this idea that no one really cares to sit back and listen to an older person talk about the life he's getting ready to complete. I remember talking once to one of my grandchildren. He was about five or six and I had this burning desire to tell him about me and how, like I say, I came to be what I am. But he wasn't biting. Not that day anyway. Didn't seem as though he really cared that much for what I was telling him, or for the perspective that I could give him. This is what I most urgently wanted to convey to him: the perspective. Just to be able to look out at all those years, seventy-nine going on eighty years, and to be able to say to someone that you lived among all those years, each and every one of them, and that you have taken a little something from each one. That's kind of something, you know. It's no major accomplishment, mind you, but things have to work out pretty right before you can say that you made it to where you are now. I'm going to be eighty. That's something, isn't it? I've just about lived as many decades as most children have lived years. Some would say, like I hear the nurses here talking in the halls, that that's enough time. 'He's led a full life,' I hear 'em telling one another. I can understand what they're about. They look at an old man lying in a hospital bed just about having to be taken care of like a baby, so they can be free to make those kinds of statements. But let me tell you something. I don't care what you hear, there's not a man or woman alive, seventy-nine, eighty, one hundred and eighty, is ready to give it up. They say that I've led a full life and that it's all right for me to die. Well I ain't ready. How do you like *that?* I'm not planning to give up, 'cause what they don't know is that thing I told you about, the perspective. What they don't know is that I got a future to think about. Now, what the good Lord's got planned for me, and my soul, well, that I can't tell. But if the Lord has patience, I've got a future."

The heat in the hospital room of the Veterans Administration Hospital was oppressive. Although it was early April, spring had come to the northeastern sections of America, and still the radiator poured heat into the poorly lit, sterile room. Four men, all in their seventies or eighties, occupied beds in the corners of the room. White curtains hung on rods attached to the ceiling so that with one pull, visual privacy could be obtained for medical examinations, bed baths, urinating, and defecating. Usually the men slept, but their color and their open mouths always made me think they were dead and that no one as yet had noticed them. On each visit to the hospital I experienced this same image: the sight of a sleeping man, the sheet still and my own heart dry and quiet on believing that I had been the first to discover a corpse. Then there would be a snorting sound as he would gasp for breath, or maybe his eyes would open slightly and then close again, and I would know that he was alive. He's among the living, I would say to myself. I'm not sure I want it to come to this.

Marcus Nathaniel Simpson, however, was different from the others. He was never asleep when I visited. He always asked when I would be arriving, although I never set up times for our meetings. "That's fine," he would say, "I'll be ready for you. Make myself pretty and all." And he would. His white hair would be combed neatly and he would be sitting up in the bed, pillows supporting his back and head, the top sheet folded over his chest. He detested the night shirts that patients were required to wear. He commented on how they tied in the back, leaving exposed what he called his derriere. "At my age that's not such a pretty sight, not even for these other stiffs." We laughed together. Mr. Simpson could be remarkably humorous, and the more I laughed, the happier he seemed. My own attempts at humor never wholly pleased him. He rarely laughed at my jokes. Mostly he just responded by saying, "Yes, that's good. That is a good one."

More than anything, he was irritated by the clear plastic patient bracelet. "Not like the jewelry I used to have when I was single and strutting down the streets of New York City, my friend. Oh, I had the items then. I had several pocket watches, absolutely beautiful, gold, beautiful things. They could sit in a museum now. I gave them away to people in my family. You know how it is, you're part of that generation. Every generation does the same thing. After getting themselves in a position where all they think about is their own age, or the day of their lives that means something to them, they suddenly get all interested in the past. So they start changing over their houses and their clothes and their cars so they can recapture the past, have it for themselves. Then my generation, the grandparents, we get a moment of glory too. Everybody comes running to us for any old furniture we might have saved from a long time ago, or like I was telling you, for the antique jewelry. I paid, I'll bet, less than fifteen dollars for these three gold watches I had. Had them cleaned every six months. I had them running like little machines. You won't find pieces that good any more. Can't afford to manufacture time pieces that good. No one could pay for them; it'd be too expensive. So, like a fool I gave them away. Soon after, the styles changed again so now they're doing something else. Now it's modern things, and they don't know what to do with all that furniture and jewelry. Ladies aren't wearing that kind of jewelry any more so that's the end of that. Leaves me with nothing else but this little bracelet." His left hand fingered the bracelet, his nails clicking on its edges. "Something I've always wondered is whether they take this off a fellow when he dies, or do they bury it with him? I suppose they must. Nothing else you could do with it. It's not like a gold watch. Reminds me, you know, when we used to grab everything we could off dead soldiers. Theirs, ours, it didn't make any difference then. We were anxious to get out, and, being young, we believed we had something coming to us. So we would grab a man's watch or see if he had some money in his pocket. Medals too. You could usually make a little money if he had some medals, you know, sell them to someone. That was a long time ago."

"When exactly was it?" I asked. A wondrous and peculiar feeling forms when speaking to someone who has been a part of time one thinks of merely as history.

"Let's see." He began his calculation. "Must have been 1920. Yup, I was in France in 1920. Like everyone, I enlisted thinking that lots of folks would never be coming home but that somehow the good Lord would spare me, at least long enough to make it back. That's one of the interesting things about being young and living through a war the way we did. I used to think it wouldn't happen to me.

Probably everyone thought that. But you know, I used to look out at lots of these chaps I served with and could almost point to the ones I felt would never make it. I kind of knew I would. This is not to say that we didn't have quite a few scrapes with death. I could tell you stories raise the bumps up on your arms *and* legs. But I won't. Get my friends here too excited. They might pop off if I told them what I know." He made an expression that communicated his hope that the others in the room had not heard him. We scanned the sleeping faces of the men and studied the withered hands that lay so still on top of the sheets. The men slept on their backs, their sunken cheeks and thin lips barely revealing that air was passing in and out of their mouths.

"Say," he interrupted my inspection, "close that damn curtain will you. This room is like an old hospital ward, men lying here ready to die. Makes a man think he's ready for the grave. You'd think they were preparing me for the whole thing. What they should be doing is putting us in rooms with younger people so we could see some life around us. Here, all you see is yourself; it's like what we used to see in Europe. All the carts and flat cars pulling away hundreds of thousands of dead men. When they're dead, you don't go thinking he's an American and he's a German. You do when they're alive, but when they're dead you weep for them all, ally or enemy. You weep for them because they're not soldiers any more; they're dead men. Still men though. Look there at that poor chap. He's been that way, must be four months now. I never heard him talk to no one. Every day I look over at him and think to myself, that's it. He didn't get through last night. But there he is." He beckoned to me to close the curtain. "All the way around so that we can be alone. A man may be old but that doesn't mean he can't have some privacy. A man's got to learn that being alone is the most important part of leading his life. I used to think I could never be alone. Always went out and got as many friends as I could. I used to think, when I was pushing sixty, twenty years ago. . . Now that's a little hard to believe, even for me. Where were you twenty years ago? Born yet?"

"Sure."

"Well, just born maybe."

"More than that."

"I see. I see. Well, I used to think at sixty that it would be better to have lots of friends, particularly when you get older. But I knew all the time there would come a day when I would be all by myself—you know, that I would have to make it by myself, 'cause everybody would be gone. That's a funny thing about growing old. You try to find things, people, that will comfort you, but you can't do it no matter how hard you work at it. And the funny thing is that if you live too long, all the people that are supposed to be staying around with you die off. So there you are as lonesome as the day you were born. God has it figured, you see, that a fellow should die about sixty, maybe seventy, but no longer. Country doesn't know what to do with men like me. They get up to this age and it starts to bother them. It's like everyone is saying, we'd like them to be healthy, but not *too* healthy. They have to somehow put the seed of death in us." He smiled weakly at me. "Maybe they figure that if they let us grow too old we'll forget we're supposed to die. I'll bet maybe that's it. Don't misunderstand me. I'm not about to go reasoning that they want to kill us off like a pack of old buffalo. It's just that they're always talking so much about people taking responsibility for themselves, so I was wondering whether one of those high and mighty responsibilities is making sure we don't live too long. You think I might have a good idea there?"

"Well, I don't know. You might at that."

The hospital ward was unbearably hot. With the curtain drawn, the little bit of air that moved in the room seemed to be unavailable to us. We spoke in a space just large enough for his bed, my chair, and the night table on which he placed a water glass and a radio.

"You know, dying's a peculiar thing. Like I was telling you, one of the biggest problems I used to have during the war was that I might die *there*. That fact seemed to gnaw at me. It was like a pain you get when you're hungry and you can't find ways of getting rid of it. I always used to ask God, please if I have to die in this war, then let me at least be wounded over here and die over there. Please do that for me, I'd say to Him. I don't want to die in some foreign place. Although, you know, I liked it quite a bit in France. I saw all the little towns no one's heard of. You get a chance to have a bit of fun, even in war, funny as it sounds.

"I got married to my wife a few years before I went to Europe. But you have to forget lots of things when you go off to war. Sometimes you find yourself in a town with little to do. Men are resting up and you find some girls. We had many a good time, even though most of us were frightened most of the time. Any minute a bomb would go off or someone would fire a gun. But we had laughs. You lead your life right, you can have some good times being in the army."

Marcus Simpson spoke to me that one afternoon almost as a reporter. He was neither daydreaming nor communicating a wish that each word should pull him back farther and farther into time. He was merely reciting facts of his life, facts that happen to have had their origin fifty years before.

On another afternoon, with the rain coming down hard outside and the two of us hidden from the other patients by the curtain sheet, Mr. Simpson spoke of the walks he had taken and of something he had experienced just before entering the hospital this last time.

"Go on, sit down," he began. "I've been thinking of things to tell you. And take that raincoat off or you'll be a patient in this place before the night comes." I struggled with my coat in the cramped space, making certain no water fell on his bed. "Go on, go on, get it off, don't worry about me. A little rain water would be quite a treat for someone who never gets to go outside now." Finally I was settled in my chair. "I miss that, you know," he was saying.

"What is that?" I inquired.

"Oh, just going outside and being a part of the weather. I miss my walks. I used to walk several miles every day. Walking is an amazing activity if you ever stop to think of it. When you get to be my age, you think about things like walking. You think about breathing, too. They're what they call involuntary actions. You can't make yourself walk. Tell yourself to walk and you'll fall flat on your face. But what I think about is that walking and running are the ways you can tell time is moving. As a young man I never walked. Never had the time to be wasting on something as foolish as walking. I ran everywhere. I used to run errands for my mother. I was one of the youngest, only one younger, my sister Helen; so I used to run to the store for my family. I mean run, too, fast as I could go. In those days, no one got anywhere very fast. I remember riding the carts. Horses used to pull us through the streets, and we would climb up on the old rigs and have the peddlers pull us all over town. When we got tired of it, we'd jump off. The best way to get home was to walk. So we ran and walked. In the last couple of years I couldn't walk faster than a baby could crawl. Getting old means slowing up. They're right when they say that people get old and cultures pass them by. Speed is the thing that gets me. Everything happens fast. I can't keep up. Well, not any more, but I used to be able to. What always got me was that you had to live life too fast. I couldn't go that fast. A person doesn't have to live that way. There's plenty of time.

"'Course all of this is pretty damn easy for me to say. I was once as much in a hurry as everyone else is today. I see the boys playing in the park near my old house. I've been walking there every day now for almost twenty-five years. In the good weather, even on some nasty days. It's nasty out today, isn't it?"

"Yes, it sure is," I answered him.

"Well, on the days that the weather permitted, I'd go down and see what the people might be doing in the park. For a long time the town didn't have a rail built there, and getting down this flight of stairs they had took quite a bit of maneuvering for me. But they got a selectman come into office over there and he didn't forget the old men. Put a simple rail in and it made my life a whole lot easier. Well sir, now I come to what I've been meaning to tell you. When I walk in that park and see all the children playing their games, I think to myself that I can still do those things. I feel that my body could still be cranked up to play like they play or push some little angel on the swings. The sad part is that I have to tell myself that I can't do it. I'll be there studying, watching, thinking I can throw a ball still. I *was* quite the athlete in my day. Then I tell myself, you're an old man, Simpson. You couldn't even stoop over and get the ball if it came your way. Who you trying to kid? Certainly not the person who handles time. But there I'll be, thinking I could just climb back into my twenty-year-old ways.

"I suppose this all sounds foolish to someone as young as you, but you'll see when your time comes. There's a voice in you that says you aren't young any more and don't go thinking you are. 'Course, what intrigues me about this whole matter is, just how old is that voice inside me? Oftentimes it sure does sound as though it's mighty young to be staying around inside of me." He looked at me quizzically.

Marcus Simpson appeared younger today than he had ten days before. A rich pink color was seeping back into his skin, and the purples, blues and greens I previously detected around his eyes and mouth had all but vanished. Today, as a portrait painter, I would have selected yellows and oranges and a peach color, perhaps, to represent his glow. What is more, he seemed to move with greater ease. The stiffness that was part of a lingering arthritis seemed less painful.

"It's funny," he was saying, "when you're a boy, you have a set of ideas that you think should determine what you'll be. That voice is there, too. Even as a boy, you can hear it. Then it's the voice of a man, a man you can look up to. In the middle years that voice goes away, probably because you're either doing what you set about to do, or maybe because it's getting too late to make much of a difference any more. During that middle time, a man is unable to hear the voice that sort of guided him through his childhood. He hears the sound of his own voice now. A man talks to himself about business, about how he's doing, and what he's up to doing in the future. But later on, the voice appears again, and now you come to realize that it's most surely the voice of your childhood and your early manhood speaking to you. Can't be anyone else. But you have passed it all by. Everyone is gone and you're kind of like the last person on the earth. I can tell how the young fellows look at me at the park. What they'd like to do is help me across the street or find a clean place for me to sit down. But I'm watching them real close and wondering whether I might still be able to run around the way they do, all snappy and peppy and eager to get out of breath. Oh, it's good to be young like that and be able to get your body to respond to such simple desires. I don't ask for much. It's painful to have all the pain. That's kind of a joke, isn't it?

"Pains me to have the pain. My body could take it. Fellow can sustain a lot of pain in his years. But it gets me down when every day I wake up I got a pain here or pain there. The real pain of living is up here." He placed the index finger of his right hand slowly on his temple. Accumulations of melanin showed on the back of his hand. The skin there seemed dangerously thin. "Nowhere else can get to hurting as right up here." He continued tapping at the side of his forehead. I looked at the white whiskers on his face. "You ask all those psychiatrists. You can bet they will agree with me about that." Finally his hand fell back on the sheet. "That's what troubles me— knowing I can't do what the younger fellows are doing when it seems like no time at all has elapsed from when I could go out and do anything with the best of 'em." He paused for a moment as if to listen for an echo of his words. "Look at me bragging. Don't go listening to me when I talk like this. I can hear what I'm saying, you know, just as plain as you can. Right now I'm hearing an old man, sounding like I was as good and fresh as these young boys today. Damn silly of me." He was shaking his head vigorously. "Now I'm going to let you in on a surprise."

I cannot express exactly the feeling that intermittently overtakes one in hospital wards with older patients. It is a complicated feeling, comprised of many sentiments and associations. One thinks, for example, of his own involvements with older family members—a grandfather perhaps—except that in my own case I cannot remember either of my grandfathers. One imagines, perhaps, speaking with a messenger of time, someone who brings not only wisdom but the secret of life. But with this desire to meet with the ancient ones among us, there remains a belief that someday one will be given something by these older ones, a legacy, an inheritance. When Marcus Simpson mentioned his surprise, therefore, I could not push aside the thought that there might be a sum of money, no grandiose estate, mind you, that he would, upon his end, sign over to me.

"A surprise?" I questioned, moving a bit in my chair. "What have you got?"

"I've got a little piece of information I think you might like to hear about."

"Go ahead," I replied with anticipation.

"Well sir, this may come as a shock but I don't mind telling you that there's life in me. Just this morning a nurse came in to take a look at us. She's new. I would know if she had been in here before. Well, she comes in looking pretty fair for a rainy day. When she came by my bed to fix it up, I gave her arm a little squeeze, and I could tell that just a bit more of life was hanging around in my old frame. Not a lot, but enough to know they haven't switched the lights off yet." I must have had a grin on my face. "Now don't go laughing, young man that you are. Don't you go acting like everybody else. We got a set of attitudes in this country about older people that could make a man sick. Everyone always laughing after the ways of older people. Like, you'll see in the newspaper an old man like me goes and marries. Damn, you see how they got me calling people my age old. We aren't old. One hundred, that's old. Seventy-nine doesn't have to be so old. Well, they got this old couple photographed and written up on the front page and you can tell everyone under sixty's going to snicker at the sight of it. Everybody's eager to ask them the same question: Do they or don't they make it in the shed?"

"In the what?"

"In the shed. The shed. Now, I got to make it graphic for you?"

"No. I think I got it."

"Well, I certainly hope so. You aren't much of a young fellow at all if an old coot like me has to go explaining what a man's going to do with his bride. But that's my point, don't you see. The newspapers, because they never have taken the time to learn about older people, don't have the slightest idea of what older people are capable of doing. And I got a mind to tell that city miss one of these mornings just what the old man feels capable of doing. No sir. The light's not gone out yet. If I didn't know it for sure the last time you were here, then I know it for sure today. That was my little surprise. I thought you would be interested to know this, what with your work and all." He pointed his finger at me. "I knew it."

"You were right," I said.

"Huh? Was I right?"

"You were absolutely right."

"See. Not such bad shape for an old man, huh?"

"Pretty darn good shape, I'd say."

"Well, I wouldn't go using any superlatives about a fellow in my condition. The important thing is that, well, there are two important things about your life when you've lived as many years as I have. First thing is you keep telling yourself that for a man your age I seem mighty old, but to a fellow five or ten years older than myself, I seem mighty young. I feel a whole lot younger when I see a man hobbling along the street. You've seen the crutches they have now for old people? The one that looks like a little balcony railing made out of aluminum so's it can be light. Then he hobbles along and keeps lifting this ugly damn thing and putting it down in front so he won't fall over. Most unsightly thing I ever saw. I have half a mind to suggest they put you in a basket and drag you around. Well now, that's one thing. That's why one needs a lot of people around him so's he can compare himself and see how he looks in their reflection."

"I do that same thing. You like to be with people older than yourself 'cause it makes you feel young." He rarely looked at me when I spoke. Although my words reached him he never acknowledged me as their source.

"That's it, you've got it. That's the feeling. I remember having it too when I was your age. You make me remember a whole lot of things."

"Is that good, Mr. Simpson?"

"Yes, indeed it is. Memory's an important part of living. Did you ever stop to consider the wonder of the mind that it's able to grab for things that happened long ago? That someone can say something like you just did, and the sound of your words makes my mind dance with memories? Not all of them good, mind you, but they come back as fast as you can imagine them. Very little of our lives goes away for good, you know, and that's a comfort to a man. That's a comfort to an old man who has known loneliness. No matter what they try to do to you, and in a hospital like this they can serve up all sorts of indignities. They can take away fifty feet of your intestines, but they can't take fifty seconds of your memory. 'Course memory doesn't stay the way it was. It begins to slip, so quietly you hardly know it's been retreating on you.

"Take, for example, the time I was on one of my walks. I found myself walking along a path that I hadn't stumbled on before. There were some benches there and being pretty tired and all, I decided I would take a little rest. A while later a man sits down. I take little notice of him. Then he asks me for the time and I give it to him. When I see his face the first time, I know that I know him. I look hard at him, see, the way I'm looking at you now, but I can't get his name. I got all sorts of names coming into my head but none of them's correct. This is no easy task for an old man, because I have to go through all the years to see what names my brain comes up with. And like I told you that other time—do you remember?—the pain of getting old is the pain of seeing your body barely able to keep up with the passing times. All of a sudden you can't do things any more, and you ask yourself, now when did my eyes get so bad? When did my hearing go? Would be nice every once in a while to ask a doctor these questions. When did my mind start to fade, and when did my body just go kaput?" He looked at me, letting his eyes fall on my forehead and hairline. "Yessir, don't let anybody tell you that this aging thing's an easy matter. Hardest damn thing I've ever had to sustain, if that doesn't seem strange to be saying. I'm living along each day, just like you, although at your age a man is making all sorts of plans. Aren't you now?"

"Yes, I guess I do." I was continually impressed by his accurate recollections of what he had previously disclosed to me.

"Of course you are. A person your age must plan. That's what life's all about. If I knew at your age what I know now, how frivolous so many plans turn out to be, I'd still be planning, making my preparations as best I could. What you find out is that living day by day only seems possible when you're getting on in years. You don't go telling a man to live second by second when he's in the middle of all sorts of doings. He'll throw you out just for suggesting it, probably. Fact is, as a child, you're not thinking about time in these ways, and so there's no sense in talking seriously to the little ones.

"Now, I like to think of the middle years as the years when that voice inside a man stops, or can't be heard, and everything about time becomes a sort of routine. There are days when you'll ask yourself, 'What am I doing this for?' Or, 'Is this what it's all adding up to, 'cause I didn't figure it that way.' But most of these adjustments don't add up to too much. At the time, of course, you're all wrapped up in yourself. But when a man reaches my age, these days are mighty important, and they're mixed in with little specks of fear. I find myself lying here hoping I might see you again. But I don't want to see you too soon if that means that time is moving on too quickly. Let him wait, I tell myself. Let time wait a moment too." He stopped talking and closed his eyes. I remember sitting upright in the chair, a shock of fright exploding in my heart. Then, as God knows what thoughts went through my mind, his eyes opened.

"I think maybe I'll be going now, Mr. Simpson," I said.

"Well, I think I might appreciate that. I suddenly felt very tired there. I don't want to be rude and fall asleep on you. You'll be thinking that I'm not worth the trouble of visiting. A man has to be a good host if he wants his visitors to come again."

"I'll be back," I said.

"I know you will. When a man gets to be my age he can tell in the wink of an eye whether another man's honest or not. I know you'll come back, 'cause anyway, I ain't through yet."

About two weeks passed before I made my next trip to the hospital. I never called in advance of coming despite the fact that the Veterans Hospital was a good drive from my home. Upon reaching his room, I opened the door slowly, seeing first the men on the right side of the room. Marcus Simpson's bed was hidden by the open door. I peered around, noticing that the man occupying the bed next to Mr. Simpson's bed was not the same person who had occupied it on my previous visits. Then, as a feeling of horror overtook me, I saw that Marcus Simpson's bed was empty, the sheets and light blue blanket neatly arranged on it.

I ran to the nurse in charge of the ward.

"Do you know a Mr. Marcus Simpson from in there?" I pointed toward his room.

"Yes. Indeed we do. He's fine. Don't worry yourself so. You'll find him up in the pavilion area." Thank God, I thought. Simpson, you old devil, you're fooling them and I'm rooting for you, but don't play tricks.

On this one visit the old gentleman sat upright in a wheel chair. He looked, if it does not sound paradoxical, better, healthier, and yet older. His color was good, his mind fresh and alert, but his being, somehow, seemed older.

"Two weeks is two years in the life of a man his age," a doctor had told me once when we spoke about Mr. Simpson's condition, the nature of the diseases that lived inside his organs, and, of course, his chances. "You can't go using your normal way of delineating time," the doctor had said, causing me to think about my own research interests rather than the fate of a man with whom I was now involved. "Calculations go out the window in geriatric cases. You don't make predictions. I wouldn't tell you what you might face in a patient in an afternoon after checking him in the morning. Hours and days don't make sense to people like your friend there when sixty-five, seventy get behind them. I've seen many cases I was about to release, figuring they were doing so well that they might as well have a couple extra days with their family, friends, whatever, pop off that night. You just can't make any sense of it. It's like, what? Well, picture yourself in an airplane in which you're running low on fuel. But the catch is your gas gauge won't work. And you can't land. So when do you run out? Sooner or later you will. That's about what goes on."

Mr. Simpson had plenty of gas left on that Wednesday afternoon in the seventh-floor solarium. "Well, well, where have you been?" he greeted me. "Thought I might not be seeing you any longer. Thought maybe you found someone down there interested you a bit more than I did."

"No sir. I just can't keep up with you. Going to your room and all, and having to track you down, finding you here with the hoi polloi taking in the sun."

He smiled, looking out at the grounds below and the highway that led back in the direction of my own home. "Well, you get to be a busy man, it's not always easy to keep up."

"Hoi polloi," I repeated.

"Now don't go giving me that stuff. Let's get to work. I've been doing some thinking and any minute they'll be sending one of these old biddies in here to take me back. Can't always tell whether they're running a hospital here or a state penitentiary. Lifers is more than what they treat us like. Lifers waiting for some parole board group to think about their case. Guess you might say they got me on a kind of parole situation today."

"Well, I'm not sure I'd. . . ."

"Now then," he interrupted me, taking no notice of my words, "let me tell you some thoughts I've figured would interest you."

"Okay, I'm ready."

"Bet you thought the worst when you saw my bed empty though." He laughed, and with pain, turned to look at me.

"I had a couple of thoughts," I replied. "Thought maybe you had gotten out of this prison and were out messing around somewhere."

"Only messing around I'm about to do is you know where." Even as he spoke his thoughts were already changing. "A man can do a lot of thinking in a place like this," he said. "Fact is, it's a shame anyone has it end for him sudden, like when he's perfectly healthy one day and then pops off in his sleep one night. Or taking a nap, like what happened to a friend of mine. In 1958 we had lunch together in the neighborhood. Sat together for a couple of hours and talked about what men, I suppose, talk about. Then the next morning I hear from the fellow lived upstairs that the man is dead. Took a nap that afternoon and never got off the couch. Funny, too, because I heard the sirens that afternoon. I realized it must have been the police coming to take him away. He was a young man.

"But what I was starting to tell you is that a stay in a hospital, even with these rotten illnesses your body finds for itself when you get old, helps. You get a chance to think about things, take a little stock, see what's been disappearing on you before you even took notice of it.

"I was thinking what it's like. When I was a boy, a bunch of us used to swim over here at one of these public pools. It wasn't really a pool like they have for the children now. What they did was let some water pour all around so the people could get cooled off in the summer. Well, I remember—funny what sticks out in your memory—thinking all sorts of things before going into the water. I was preparing myself, like a woman will go preparing herself in front of a mirror before she comes into some important room. Well sir, that's the way it should be. A person has to have a chance to prepare himself for the whole thing." For a moment he sat in silence. I looked quickly at the few other patients in the solarium. No one spoke to anyone. Five men sat in chairs or wheel chairs looking out the window, their bodies moving only slightly in rhythm with their breathing. I hoped Mr. Simpson would not notice my glancing about the room.

"Funny though," he was saying, "about that memory thing. I thought I saw snow fall this morning. Didn't think it could really be happening what with the sun shining so nicely the way it is, but something was falling. Well, I was reflecting on the way seasons come and go. They're like a clock, you know. You wouldn't want to set your watch by the changes in the world outside, but they come and go. Well, I was thinking about the seasons and the way certain memories come back to a man as though they were attached to the seasons. A man my age gets to the point where he can fetch old memories merely by telling himself to think of the winter or the spring. So this morning when I saw what I was certain looked like snow, a thought came back. I must have been six or seven at the time. A bunch of boys were out somewhere playing. Must have been November. We're out there doing what boys do, when a few snow flakes start to come down. I remember one of the boys saying, 'Here comes the snow,' you know, 'winter is about to begin.' But I up and say, mind you now I'm this little shaver, 'That's not really the beginning of winter. God's just sweeping out last year's snow so that he can get ready for this year's snow.'" I smiled, but as usual he took little notice of me. "I can't have been more than six, seven at the time. Strange isn't it, where a boy will find the things he says? We got a little bit of religion when I was a child. My sisters and brothers and I used to go to church pretty regularly. Still, where does someone get an idea like that?"

"Mr. Simpson," I began, "when you look across your life, you know, with the perspective that you have. . . ."

"Ah, you remember my words to you, do you?"

"Yes, indeed I do."

"Yes. Perspective. That was my word to you."

"It absolutely was. Well, do you see a connection between now and then? Can you see yourself as a little boy? I mean, do you know that little boy is you or do. . . ?"

"That's a good question. It gets right to the heart of it. That's the amazing thing, you see, what God does for us. I see a little boy, like I was just telling you. Could be any little boy. I know that it's me, but I'm not really certain. I just see these images of life, years here, years there, but I can't even remember what I looked like, although I think if you showed me a snapshot of myself in a crowd I could pick myself out. Even so, I got to go telling myself, that's me." He pointed to an imaginary photograph on his lap. "That fella there in the fourth row is me. Pretty tough looking rascal, wasn't I? But the ladies used to think I wasn't too bad. Never was very big. I used to think about how much bigger I might get. I remember a fellow telling me that after seventeen or eighteen a man doesn't grow any more. I didn't believe him at first. Guess I didn't want to. I kept having this dream that I'd be a big tall fellow, but I never made it. Now you're a pretty tall fellow, even for this day and age?"

"I guess I'm bigger than average, though they have these basketball players who make me look like a shrimp." There's no getting around the self-consciousness a healthy person feels in a hospital.

"Well now, you see you have one on me, because I've seen you sitting and standing; but you've only seen me sitting or lying down. You wouldn't know how tall I am." He lifted the covers off his lap to expose his legs. "You can see that I do have legs. Could you give me a hand here, please." The blanket had slipped and was caught between the side of the chair and the wheel. I covered his legs, trying to imagine what his height, as a young man, might have been. "You know," he continued, "one of the saddest days of my life was looking into a mirror one morning, must have been fifteen years ago. I must have been dreaming that night, because I suddenly expected to see the face of a young man. Well, you can imagine what I saw. I stood there looking at myself. Then, of course, I got self-conscious and started looking around behind me to see if there might be someone else in the room. Then I had to look again. I suppose I was hoping I was wrong the first time.

"I think I told you once about feeling inside that you're young, or younger than you are. That there's one of the strangest feelings about being old. Part of you believes you're young; part of you is getting ready to get used to the idea that you're old. 'Course the word 'old' doesn't mean a damn thing. When I hear the men on television talking about old timers or an old people's home, I still have the ideas come into my head I had when I was a boy. It's always them, I think about. Crippled old people barely able to walk, ready for a grave. Then one day I look into the mirror, and there's this old man looking back at me. Then the half of you that thinks he's still young has to learn from the other half that you got to start getting ready to be old. Damndest sort of procedure I think anyone could invent!"

I watched him move slightly in the chair. I could see the uncomfortable way he sat, but even with the physical pain he endured, pain that he never complained about, I felt it was the ideas that crossed his mind that caused him this last discomfort.

"These very special moments are the moments when one gets the slightest inkling about the passage of time. What happens is like that business with the mirror. You look in and see someone and know something has happened. Time has gone by, you say to yourself. Like when you meet a friend in the street. I used to see my friends at the pharmacy—old geezers getting together and reminiscing. Then maybe you'd spot someone you hadn't seen in years. 'How many years has it been? Thirty-five. My God, my God. Where has the time gone?' That's the part of time I was referring to. Where has the time gone, you ask? But every once in a while at my age you can feel time passing. You can hear it too." He leaned toward me as if to authenticate his words. "You can hear the time passing. Shhh, it goes. You don't need to meet up with old school friends then, or look into any mirrors. You can tell you're moving on. The young can't know this. But after a certain age. . . You know that expression, your days are numbered, or are you too young for that one?"

"No," I laughed, "I'm not too young for that one."

"Well sir, I don't think it means exactly what people think it means. What it really means is that for a small bit of your life you can be aware of time passing; and now here's my punch line, time passing and taking you with it. You thought I was going to say time passing me by. But that's the feeling all of us should know from when we're young. Not real young. Not children. God doesn't intend for them to know anything about what we're talking about today. Your age, I mean. *You* know the feeling of time passing you by, I'll bet?"

"Yes I do. Every day."

"See there, that's what I mean. And that's because you have all these plans and ideas picked out for yourself. So much to be done, you think, and not enough hours in the day."

"That's the feeling all right."

"Oh yes, yes it is, my friend. That is most certainly the feeling we human beings, and only human beings, know so well. Animals are spared of this. They can't know these feelings, for it takes the brain of a human being to know this. Time passing you by is very different from time passing and taking you with it. When you are young, you see, no matter how hard you think about it, and no matter what they tell you in church, you cannot imagine death. Oh, you use words to describe something, but you cannot get to it. This I know for a fact. Fellow my age has a few things in his head he knows as facts. It doesn't matter how many times a year you say I'm wasting time I can never get back, because you're still of a mind to believe that time is not really passing. You don't feel anything that tells you it's passing. But now, as I get old, I *know* it's passing. Like I told you. I can hear it. Time tells me this is it, and I'm beginning to believe this time it is."

"Are you," I tried to form my question. "I mean, do you have feelings of being scared?"

"Sometimes I do, sometimes I don't. When I do, it's very scary indeed, far more scary than it used to be. Remember now, I'm living a life where I see people leaving this earth every day. Take a person like yourself, I'll bet you never turn to the obituary page in your newspaper?"

"Yes. You're right. Very rarely."

"Well, of course. No reason for you to be seeing who's dying on a particular morning. Me, I used to study it like a fellow studying the stock market or the real estate pages. You know what I'm looking for, don't you?"

"Yes, I think I do," I replied.

"Well now," he went on, "here is what you might call a surprise. Even with the knowledge of what's happening to a person, I still go on dreaming about my future and making all sorts of plans. Nothing foolish, of course, and everything resting on the chance that I'll be getting out of this prison one of these days. I have a notion it might be a good idea to straighten up my room and see whether I can't find some friends I know must be in the city, someplace. Lots of them are gone, I'm sure, but I think there would be a lot left, all probably wondering many of the things I'm wondering about. So you see, I'm thinking quite a bit about the future. That doesn't stop. Don't let anyone tell you that it does. You know, though, I've been realizing something peculiar about what I've been thinking. You've been getting to me, I believe, even though you sit so quiet most of the time."

"How's that, Mr. Simpson?"

"Well, it's not easy for me to understand, but just by doing all this talking, a man gets to see the way his mind is working. I've been watching myself, like a sentry, and come up with several items. Fascinating the way a mind works, even an old, worn out mind like mine."

"I'd hardly say it was worn out," I offered.

"Would you permit me to say, then, that it is wearing out?"

"I'm not so sure I'd even accept that, Mr. Simpson."

"Well then, you are a difficult man. I might need a lawyer to get through to you."

As he spoke, I reflected on the stillness of his hospital room and the solarium. Our conversations were the only ripples flowing into the quiet air. Our voices were like images on film playing on a screen of space. The normal bustle of hospitals was absent, particularly in the solarium, a fact that made the topic of our exchange that much more delineated. Perhaps my thoughts were merely the twisted elaborations of a question that could hardly be enunciated. Was this the last time?

Marcus Simpson's legs moved under the blanket. He had attempted to lift one foot up and place it on the other, but his muscles could not sustain the effort. He stared off into the room, and as I imagined his body to be weakening, his voice came back as strong as ever.

"Well sir, if you can believe it, my future unravels before me. It brings to me all sorts of visions. I see it as a library filled with books and ideas. I need only grab a book off a shelf and my future will be there. The gist of it seems rather meager. The only thing I can be sure of is that it's there, out there somewhere where the seasons turn, come and go. I know that I shall see the seasons again, but I have a strange feeling, too, which my mind has caught. When a man gets my age, he can't see the future with people any older than himself, or not too much older than himself at any rate. If I was to imagine people too much older I'd be looking in the province of Saint Peter himself, wouldn't I?"

"Yes, I guess you would."

"Well of course I would. I'd be doing the inspection work for the Lord." He laughed quietly. "So what your mind does for you is to make you see the future with people who are children. I see my future filled with children. I can see one boy in that world I believe is me. I say believe because I can't tell for certain. Sometimes I see his face and I believe it is me. I'm almost certain of that. Then I see my grandson's face. It's a different face from the first one. They change over right before my eyes. 'Course my eyes are closed and I'm using my inside eyes. Those are the good eyes for they never go blind.

"Now, getting back to the children, I can tell you that my images are not the same. They're inconsistent, if you know what I mean. I was doing quite a bit of thinking on this just the other night. I got to thinking of a theory of life, kind of the way time plays itself out in the human spirit. At first, my theory went that you end up where you started. You come out as a baby and you leave as a baby, all shriveled up and helpless. But that sounded too simple. So I worked on the problem a bit more. There I was, one of those nights, I couldn't tell you what night it was, 'cause I never know in the prison here one day from another, though that might be a blessing. . . . What was I telling you?"

"Lying in bed and thinking about time. . . ."

"Yes, yes, yes. Well, I'm lying there thinking up my schemes about time, working like some important philosopher, thinking what could it all mean, and why does a person have to live the way he does, when a nurse comes into the room. For an instant there I had a desire to talk with her, explain to her what I was pushing in and out of my mind. She's a gruffy old dame, not too well liked by the boys. But I decided I better not let on anything to her because she'll just try to shut me up, call me an old man babbling on about this and that. So I shut my eyes to make her think I was sleeping. When she left I opened them again, very slowly to make sure she wasn't there. I do some of my thinking with my eyes open, too. Then I got to thinking how sad it is that I couldn't speak with her. I was doing a bit of thinking for her, too, making a decision for her, sparing her from a babbling old man. Well, maybe that's what I am doing."

"You're not at all, Mr. Simpson."

"Well, maybe I am. You might be too polite a fellow to let on."

"No, I think I'd tell you."

"It's kind of you to say these things."

"I'm eager to hear what else you were thinking." I remember thinking that the usual pleasantries, routine as they are, were absent from our meetings. Rarely did we utter good-bys, or thank-yous, or how-are-yous. With few beginnings and mild endings, we merely resumed our conversations, what we agreed was our work, doing our best to sustain a sense of continuity.

"Well now, let's see," Marcus Simpson was saying. "Future, children, not my own face. Oh yes, those thoughts I was having when that old toughy came into the room for inspection. My theory, if you can call it that, came down to me thinking that the end of life is nothing more than, well, an end of a unit. Many units together end and make it possible for a whole new generation of young people to get moving. My effect on my grandchildren, you see, is only indirect. I didn't cause them to be around. It's their parents, my children, that did that. But we're all part of a chain, a chain that holds the strands of time together, if you get what I mean. Seems a shame, though, that not all the links have that much to do with one another.

When a person in your family dies, it rubs off on you, makes you wonder what's happening, what it's all about. When one part of the chain goes away, people have to move around and see whether they can find another link somewhere to make a connection with. Sometimes you feel like you can't replace the link; sometimes you're kind of surprised how quickly you can go forgetting things. But they come back to haunt you when you're least expecting it.

"I lost my wife almost twenty years ago. Not a day goes by I don't think about her. Sometimes I think about her when she was young. Sometimes I can almost see her the day right before she died. Sometimes though, no matter what I tell my mind, I can't get myself to see her at all. Not a single part of her. It's like she's fading in and out of my memory as though maybe I don't want to think of her not being around any more. It's not easy when you're a widower. I wouldn't wish it on people. Well, I'd better not get to thinking about that when you're here. Don't want to be wasting time now, do we?"

Marcus Simpson had one last thought about the children in his future. It had something to do with time not ending up where it started, and the difference between the generations being the feelings connected with the lives of the people constituting those generations. It was an incompleted thought, one to which he never returned.

It had been six days since our meeting in the solarium. Marcus Simpson was back in his room on the fifth floor, his eightieth birthday one month away. He looked still better today, as healthy as I had ever seen him. Strangely, but thankfully, his condition was improving. The doctors were encouraged, and talk of his leaving the hospital had reached him. Our discussion about the future now seemed prophetic.

"Religion is what I had in mind to speak about today. You a religious man?" Mr. Simpson inquired of me.

"I'm not sure. I'm getting there."

"Well, a person your age is too young to know much about religion." He fingered the patient bracelet, and without looking up he muttered, "They just won't let you forget. Damn prison." He took no notice of my expression. "When you get on," he was saying, "you begin to see the logic of life. Time just unrolls like a huge rug. Over and over it keeps on coming, letting people walk on it and laying itself out for others who might be lucky enough to join the parade. For a while, it seems pretty right, the logic of time. You begin to depend on it, knowing the way some things will happen and some things won't, no matter what you try. You get to know the way evolution is going to run its course with you.

"Then, somewhere down the line, you get to thinking that the logic isn't making you happy. Fact is, you'd like to make a big dent in time. You know, kick it in like a young punk kicking in the fender of someone's car." He made the most forceful gesture with his fist I had seen him attempt. "That's the time, it seems to me, a man wants religion. He wants the scenes of his life to shift so no one can predict what's going to happen. Sometimes, you see, you count on knowing that the sun's going to come up just where it always has. But sometimes you'd like it to forget a day or two. When I needed a bit of religion was when I wanted things to get less logical. I didn't want explanations any more. I wanted beginnings where the ends were and ends where the beginnings were. And just the fact that nothing would make sense was what I hoped to achieve.

"But right there I have said the word. Hope is the word you hunt for in religion. Without it you're like a horse without a rider, going it all by yourself. Being by yourself is not the worst fate a man has to face. But being by yourself without hope, well, that's a disease without a cure.

"But you know, you hunt for religion in the same way you hunted for things when you were a little boy. Where did mommy hide the cookies? you ask yourself. Whole thing about life is that all the essentials, all the things you need, are right there in front of you. You can't look back with longing. A man that looks back is a man about to trip and fall flat on his face. No sir, you look forward and try to find the hopeful things in this world." Then looking straight at me, so straight in fact that I felt a quiver of fright, he said: "Never forget I taught you that. You will be thankful one day you've got that lesson under your belt. But many, many years will have to go by before any of what I'm talking about could make any sense. Someday your time will come and you just might be seeking religion for some of the same reasons that drove me to find it."

For one reason or another his words made me reflect on the hundreds of questions I had not asked. I knew so little about this man, even now after hours of visiting with him. Where he was born, his parents' interests and occupations, the nature of his family, his travels, his own education and career, remained mysteries to me. All of the stock sociological inquiries to which I invariably pay attention were missing from my notes. It was about life that he spoke, his life to be sure, and the content somehow was there without being there.

"Is it getting late?" Mr. Simpson asked. Neither of us had spoken for five minutes. Nothing seemed wrong. He had merely sat there in his bed, eyes open, heart beating regularly as far as I could discern from the slight movements of his chest under his hospital gown.

"Yes, it probably is about four. I don't wear a watch," I answered.

"You don't, huh? Now, isn't that something. My, when I was your age I always had to know exactly what the time was. Couldn't go anywhere without one of my gold watches. I had several of them. They were beautiful pieces. Gave them all away to young people in my family, young people not too much younger than you are, I'll bet. It was the wrong thing to have done. People your age take a fancy to something one day, and the next day they move on to something else leaving everything, all the valuable things and advice you gave them, behind. A man my age can't afford to be that way. If anyone gave me something now, you can be sure I would care for it all my life, or, well, as long as I was able to. Not the young, though. Those gold watches were museum pieces. You know anything about antiques you would know what their value might be today. Museum pieces. Many a person would be pleased just to stand in front of a case looking down at my watches. They would be working, too, telling time exactly like they used to. That is, if anybody ever took the time to care for all the things that came out of the past, instead of getting new things which capture their interest one day, and then the next day getting interested in something else. Sometimes I think we do the same thing with people. Protect them, care for them a while, and then, when they get to be old or boring or tired, drop them and find someone else. Maybe before long they'll pass a law saying that when you got something you got to protect it so you can't give it away no matter how much anyone wants it." Then he was silent again. He barely moved, his gray eyes looked straight ahead, his eyebrows raised slightly as a thought came into his head, then dropped down again. I could hear the sounds of our breathing. The three other men in the room slept. Occasionally I heard one of them gasp for air, then everything was silent.

"I get to feeling tired this time of day," he said. "Don't have anywhere near the energy I used to. Seems to disappear so quickly. Almost feel it running out of my system. I used to have so much. I was never a big fellow, but I was always strong. Wasn't a thing a man could lift I couldn't lift right along with him. Lifted boxes ten hours a day once, six days a week, and earned less than two dollars. Can you imagine? Today a man would earn himself a lot of money, time and a half and all. I tell myself I was born at the wrong time. Now's the time to be alive. Now's the time everything's happening, or getting ready to happen. So much exciting beginning to take shape. All the countries coming together now. Lots of things to be learned and to be thought about. Well, I'll leave the thinking to your kind. You seem to do a lot of thinking. That's probably good, too, as long as you get a little fun mixed in there."

When I departed that afternoon, Marcus Simpson held his hand out for me to shake. Remaining silent, he nodded his head as though to say, yes. I nodded back at him. His hand, I remember, barely squeezed my own. I felt the tubular outlines of the veins running along the back of his hand and recall looking closely at the spots of melanin on his hand and forearms. Before leaving the room I waved to him, but he was no longer looking at me. Then I proceeded with the ritual that accompanied each leave-taking. I repeated the last words he had spoken, just in case: ". . . a little fun mixed in there." The words were so innocuous, so meaningless and pale. "Mixed in there."

Four days later, deep into the night, Marcus Simpson passed away. A night nurse discovered him. He lay on his back, his left hand fallen on the wrist that wore the patient bracelet he so disliked. He never received my last messages to him; he never read my words, his words, really, as through me they might have been returned, somehow, to him.

The conversations with Marcus Simpson reveal a man oriented not so much to periods of time as to time itself. This is the perspective he communicates; it is an awareness of the passage of moments through life that pull and stimulate him. A philosopher of time, his credentials are derived from age and that special wisdom of older people. A social psychologist as well, he speaks of the effects of people interacting on the meaning of time, and at certain moments, we feel that death can only come to him when he is alone.

This idea about death is elaborated in his religious feelings. He speaks of the imponderables in ways that violate his own sense of logical evolution. Religion has an anti-predictable quality to it, a possibility for surprise that he cherishes. The realistic future that he guards is kept separate from a religious re-ordering of the world that he might wish to bring about. But like many of his thoughts, religion leads him to contemplate death and the natural realities of human evolution. "Hope is the word you hunt for in religion," since sentimentality and a longing for the past appear somehow to be unhealthy and spiritually unprofitable. To unite the peoples of the world, or the generations that form one's family, is an engaging fantasy for Marcus Simpson, one that clearly yields him a foundation for hope. Nevertheless, he builds this fantasy on anticipation and a resistance to retrieving his past and substituting it for the present.

We are struck in Marcus Simpson's account by the sense of helplessness that the hospital serves to create. Objects and people are taken away from him as though to prepare him for the emptiness of dying, and he, with the others, is objectified and rendered dependent and useless. In his recollections of prisoners and of survivors stripping dead soldiers of their possessions, we hear the resolution of his own sentiments about death. The present and the immediate future are transformed, as if in a dream, and appear as vignettes in his remembrances of the first great war. Powerless, soldiers are compelled to flirt with death, knowing they possess such meager control over tomorrow, or indeed over the rest of the day. Unable to handle their fear of death and the possibility of its occurring on foreign soil, they hope at least to die at home. In war, Marcus Simpson touches, moreover, on the issue of privacy—living and dying privately, free of public scrutinization and assessment. Soldiers, prisoners, and patients are taught to die. Society, he claims, "puts the seed of death in us"; no one is allowed to live too long.

It is tempting to label such strategies and reflections as forms of depersonalization and denial. He himself believes that speaking about dead persons is a form of denial, although conjuring up lost time has merit right to the end. More revealing, however, is his way of watching time pass from the simultaneous perspectives of nearness and distance. He is the man in a train car described by Heidegger, able to see the ground beneath him rushing by, the mountains miles away standing still. This is the dual perspective on which he reflects. And still, despite the reality of rejection and a biological sense of urgency that shows in his interruptions and his need to speak quickly, he envisions hope and the sense of possibility joined in the future: "The whole thing about life is that all the essentials, all the things you really need are right up there in front of you." Even at the end, the future dominates.

Time passes and takes him with it. One moment we feel him to be active and in control of time. The next moment, his words bespeak a passivity, a resignation, a surrendering to time. But time's passage, coupled with his own weakening state, transcends the lyrical and vivid images of the past, present, and future he is able to evoke. As the future, "out there . . . where the seasons turn," runs out, as he sees his own death in the progressive disappearance of the men in his room, time becomes a crystal, sharply edged and finite. The idea, finally, of a completed life stimulates in him a definition of destiny. The young, he claims, are capricious, not by dint of mischievousness or irreverence, but because they cannot comprehend the very boundaries of time; the beginning in the form of a single life, and the end.

Time, for Marcus Simpson, is also history. To relinquish personal objects is to despoil history and upset the natural rhythm of time. The young, however, cannot understand this; they cannot know the cascading images of the past and future and the revisions of these images that history forces men and women to make. The changing voice within him is more than a personification of time. It represents his efforts and aspirations, his plans and realistic appraisals of history. The tension is not between youth and old age, therefore, but between the accumulating images of one's self, and one's prior selves: "Part of you believes you're young; part of you is sort of getting ready to get used to the idea that you're old."

Day by day, Marcus Simpson's construals of time are altered, however slightly. Breathing, walking, the integration of biological and psychological systems create a beating rhythm of motion. All of this he witnesses as though undertaking his own research, and all of this he keeps from those who take care of him, and who, for their own reasons, avoid confronting the reality which is his fate. Above everything else, Marcus Simpson makes his preparations, watching himself and testing the truth of his expectations. It "pains me to have the pain," he says, momentarily hating perhaps the very nature of time. Gradually the question of how soon is translated into speed, then masked by his capacities to retrieve the past and relocate it in the present. Yet always the integrity of the past, present, and future provide him with a feeling of worth.

What does it mean, really, to say one is getting older, or feeling older? Does it mean more than being aware of one's finitude? Personal involvements and the feelings of abandonment are also part of getting older, for they generate new conceptions of the past and future, new arrangements of the pieces of life that ultimately form one's identity. One wishes to move on, remain as one is at each instant, and still look again at the past. Survival is courage, acceptance of the immutable, the quintessential endeavor.

All the while the rest of us proclaim dispassionately that old people are not yet prepared to die; that they have not accepted the facts of time. "Well," the doctor had said, "you picture yourself in an airplane in which you're running low on fuel. But the catch is your gas gauge won't work. And you can't land. So when do you run out?" Those close to death deal with it in their own way. Their image is one of mechanical failure, collapse, and descent; sure tragedy momentarily diverted, but in the end a logical result of even the best technology. They have turned away from it. And so have I, for I could not say the word, death, to Mr. Simpson or to myself in the hours during and between our conversations. Even here, in the last sentences that I will write about him, I discover that my thoughts appear in a present tense, a tense still alive, held upright by hope and a sense of the future.

PART
2

Developmental Perspectives

3
Adolescence
and the Limits of the Possible

Until youth begins to plan, the sense
of self is not complete.

G. Allport (1961, pp. 126–127)

The contemporary study of human development has
been greatly influenced by the work of two men in particular. Erik
H. Erikson has done more, perhaps, than any other psychoanalyst to
extend Freud's original insights regarding the nature of psychosexual
development. Erikson's own perspective includes a thorough consideration of a child's changing encounters with his social environment,
as the processes of maturation and education combine into a series
of decisive encounters that may profoundly shape his basic orientations. Meanwhile, Jean Piaget and his colleagues in Geneva have
transformed our understanding of the nature of intellectual growth.
They have shown us a child who is actively engaged in mastering
his environment, spontaneously reorganizing his conceptions while
coping with its changing demands. Building largely on the work of
these two men, this chapter explores the normative changes that

characterize a child's conceptions of his personal future as he grows into adolescence in Western society.

From all appearances, the new-born infant lacks any awareness whatsoever of the flow of time. He lives essentially in an ever-succeeding series of present moments. Yet, as Erikson (1959) suggested, the world that surrounds an infant is filled with recurrent temporal rhythms. These are inherent in his first experiences with the sensations of mounting tension, of delay, and of finally attaining the satisfying object. Indeed, the infant's first conditioned adaptations are likely to be to the rhythms of feeding schedules when these are imposed upon him by conscientious parents (Marquis, 1941). Lacking imagery of any sort, however, he remains as yet unaware of the qualities of time that govern his existence.

This passive responding to immediate stimulation soon merges into a lengthy period of pre-verbal play and active exploration. By the end of the first year of life, as Piaget (1952) observed in the case of his own children, actions are clearly governed by intentions, by anticipations of the effects the child may have on the objects he encounters, and by images of goals to be achieved by present actions. Many years earlier, William Stern (1924, p. 111) made a similar observation. A child's earliest ideas, he suggested, ". . . do not appear so much as memories pointing to something in the past, but as expectations directed to the future—even though only to a future immediately at hand."

One of Piaget's (1952, p. 297) observations makes this clear. His daughter was sixteen months and twelve days old at the time of the following episode:

> Jacqueline has been wrested from a game she wants to continue and placed in her playpen, from which she wants to get out. She calls, but in vain. Then she clearly expresses a certain need, although the events of the last ten minutes prove that she no longer experiences it. No sooner has she left the playpen than she indicates the game she wishes to resume! Thus, we see how

Jacqueline, knowing that a certain appeal would not free her from her confinement, has imagined a more efficacious means, foreseeing more or less clearly the sequence of actions that should result from it.

By the age of four or five, the scope of a child's activities has expanded dramatically, largely as a result of his command of language and the symbolic potentialities this provides. A child of this age, Lewin (1935, p. 173) wrote, "no longer strives solely for present things, not only has wishes that must be realized at once, but his purposes grasp toward a tomorrow." Erikson (1964a)* also recognized the emergence of the first clear rudiments of "purpose" in this period of early childhood: By the time he is four or five, a child has clearly begun to forge "a temporal perspective giving direction and focus to concerted striving" (p. 120).

These developing conceptions of time and the future nevertheless remain extremely limited. Stone and Church (1957) summarized the situation in the following terms:

For all the rapid development of ideas of past, present and future, of short and long time intervals, even at the end of the preschool period there is still no overall, consistent framework of time, but a patchwork of uncoordinated time concepts, which in no way restrict the movement of the child's thought (p. 184).

A great deal of further development will be needed before images of the future lose that freedom and become integrated into realistic conceptions of the past and the present. As Marcus Simpson mentioned, even older children have little sense of movement into the future: "God doesn't intend for them to know anything about what we are talking about today," is the way he phrased it.

*The letter following the date distinguishes between two works by the same author published in the same year, both of which are included in the List of References.

A CHILD'S FUTURE

Freud himself had little interest in a child's years between six and eleven, the time of the "Latency Period." The Oedipal conflict has been resolved through the wholesale repression of sexuality. The child has identified with the threatening parent and has acquired a Superego, now for the first time available to assist the Ego in controlling the Id. Since the dramatic intrapsychic conflicts of the earlier years have now abated, it is hardly surprising that, as Robert White (1960) observed, "Freud seems to have found the period something of a bore."

Other psychoanalysts, however, have discovered in Freudian conceptions the basis for a far richer analysis of these important years. Their essential characteristic, from this perspective, is the growth of the Ego and the increasing dominance of the reality principle in mental functioning. Part of what this development brings is a new ability to anticipate realistically and to delay present gratifications in the light of these anticipations. The primary function of the reality principle, as Freud (1911) described it, is to see to it that "a momentary pleasure, uncertain in its results, is given up, but only in order to gain in the new way an assured pleasure coming later." According to Blos (1962, p. 174), it is during the latency period that "the reality principle stabilizes the use of postponement and anticipation in the pursuit of pleasure."

The Development of Delaying Capacity

As suggested in Chapter 1, the willingness to postpone present gratifications for the sake of an imagined future depends on the degree to which these anticipatory images generate the appropriate emotional experience. If a child is to wait for a promised future reward, he must not only be able to maintain a faithful image of that distant event, but also must be able to experience now and throughout the waiting period some of the satisfaction that he will experience when that reward is finally his. Otherwise he will have no incentive for rejecting an alternative lesser reward available immediately. Is it any wonder that young children are so notoriously impatient? As Meerloo (1954, p. 29) remarked:

> The insistence of a child on getting something now, not this afternoon or tomorrow, is not mere capriciousness; the child has an instinctive fear of the unknown jungle of the future represented by tomorrow. The concept of confident anticipation is a more differentiated function of time. There is a relationship between greediness and the undeveloped function of time.

Research evidence suggests that as children grow during the early school years they are generally able to endow increasingly distant future events with a sufficient sense of reality to support a preference for delayed rewards. Melikan (1959) offered Arab children, aged five to ten, the choice between the Jordanian equivalent of two and a half cents right away or five cents to be awarded in two days. He found that the major shift to a preponderance of delayed reward choices occurred at the early age of six. When the average delay interval is one week instead of two days, however, and the choice is between a five-cent candy bar now and a ten-cent candy bar later, the proportion of delayed reward choices continues to increase with age from five until eight, and remains stable between the ages of nine and twelve (Mischel and Metzner, 1962).

As a child grows older, he regards the short-range future as less distant from the real events of the immediate present. Farnham-Diggory (1966) asked her subjects, aged seven to sixteen, to indicate on a 197-millimeter line how far away various future times seemed to be. The objective periods ranged between three hours and eighty years from the present. The distances marked off on the line were significantly related to age: The younger the child, the farther away the future seemed. In addition, the children were offered the choice between a one-cent candy bar now and a five-cent candy bar one week later. Compared with those of the same age who chose to wait for the larger reward, the children who selected the immediate alternative judged the shorter periods of six months or less to be significantly farther from the present. Future events that are subjectively closer are also endowed with a greater sense of reality through their connections with the present. It may follow that people of any age who regard future events in general as having substantial reality are more likely than others to postpone present gratifications for the sake of future rewards.

This hypothesis was explored as part of a larger study on children's conceptions of the future (Klineberg, 1968). Our respondents were forty-seven French boys aged ten to twelve who were attending boarding schools in the Paris suburbs. In an effort to measure their willingness to delay gratification, we devised two questions. The first was a hypothetical story-completion in which the children were asked to indicate whether they would prefer to receive a used bicycle now or choose to wait for a new one the following month. The second was a behavioral measure. As a way of thanking them for spending their time with us, we offered the boys a choice between a small candy bar now or a much larger one for which they would have to wait until the following week. The boys who chose the delayed reward in both questions were to be compared with those who in at least one of the two instances selected the immediate alternative.

Next, we needed a measure of the degree of reality with which our respondents endowed their images of future events. We gave them the following instructions:

Here are fourteen things that I've written down which happen to just about everybody at some point in their lives. I'd like you to tell me how old you think you will be when each of these things happens to you. Okay? How old will you be, do you think, when

—you get married
—you take a big trip
—you retire

—you can say that you have most of the things you've always
wanted

—you complete your military service

—you buy your first car

—your last child is born

—you make the most money

—your youngest child gets married

—you buy a house

—you begin your career

—you become too old to participate actively in sports

—you become a grandfather

—your first child is born

If future events are endowed with little sense of reality, a child is
likely to make rather haphazard guesses about when these events will
occur in his life. This tendency would be reflected, for example, in
estimates of his age when he is too old for sports that are either
unrealistically low (twenty-eight or thirty years of age) or unreason-
ably high (ninety years old or older). Children for whom such events
seem more real would be expected to make more moderate guesses.
The boys who consistently chose the delayed rewards did indeed make
more moderate estimates of how old they would be when these events
occur.

Later in the interviews, the children were presented with the same
fourteen future events, now written on separate cards. This time, we
asked them to arrange the events in their probable order of occurrence.
The degree of consistency between a child's ordering of the cards
in this question and the ages he had previously associated with each
event should reflect, we reasoned, the extent to which he conceives
of the future as an orderly unfolding of events in a predictable and
dependable succession. The children who chose the delayed rewards
were significantly more consistent in their two different guesses about
when these events would occur in their lives.

These results support the notion that a child's increasing capacity to postpone gratifications as he grows during the latency period reflects a developing ability to endow images of the future with a sense of reality. Short-range future events become subjectively closer to the present moment and more securely connected with present experiences. The future in general grows to be less of an "unknown jungle" as a child approaches adolescence. If past and present experiences give reason to trust the promise of future rewards, that promise and the imagined events associated with it become sufficiently compelling here and now to persuade a growing child to reject the lesser immediate alternatives with which he is presented.

If, however, a child's previous experiences have led him to believe that promised rewards will not be delivered, we would hardly be surprised to find him unwilling to reject the immediate and surer opportunities for gratification in the present. Children from broken homes or otherwise unstable family backgrounds are reported to be far less likely to choose delayed rewards (Mischel, 1958, 1961b; Vernon, 1969). This is also true of incarcerated delinquents (Mischel, 1961a) and psychotic children (Farnham-Diggory, 1966). Social-class differences are equally likely to be important in this connection, for there is much in a poor child's background to lead him to distrust promises of future rewards. These realities are explored in Chapter 6.

Distant Fantasies

The rudimentary time perspectives developed in the play activities of the preschooler expand decisively in the course of later childhood. Widening horizons and greater experience with symbolic interactions give increasing scope to a child's imagination. As he enters school, he begins to move into an elaborate social structure that defines the pattern of growing up that he will follow. From his first experiences in kindergarten and through contacts with older children, a child soon learns, as Seeley and his colleagues (1956, p. 94) suggested, "that there are innumerable stages ahead of him, in so long a vista that he can envisage neither the whole nor the end result towards which these stages are intended to move him. His horizons suddenly widen." Similarly, Erikson (1959, p. 114) observed that, as he comes to identify with older siblings and playmates, a child "begins to build up a set of expectations of what it will be like to be older and what it will feel like to have been younger."

Still, these widening temporal perspectives remain largely limited to images of later childhood and adolescence. These are the years when a child needs above all to develop a "sense of industry," when he can immerse himself in the way things work in the concrete world of the present, and when he is largely preoccupied with the challenge of coping with his changing environment (Erikson, 1950). A child is unlikely, as yet, to integrate distant anticipations of his adult years into an overarching sense of identity or an awareness of his life span as a whole. The social role of a child in Western society permits this immersion in the present, and adulthood remains subjectively far away. Furthermore, as Piaget has shown, a pre-adolescent's growing intellectual capacities have usually not yet attained the scope that later will permit him to fathom the psychologically distant, the abstract, and the hypothetical.

For Piaget, intellectual development entails a progressively more effective adaptation to the environment. In the process of coping with his surroundings, a child is forced to develop new intellectual strategies for dealing with conceptual problems. What Piaget and his colleagues have described is the way these strategies develop as they are transformed from the purely reflexive actions of the newborn infant into the logical operations of adult intelligence.

During the years from two to six, a young "pre-operational" child usually deals with the environment solely in terms of before-the-eye reality. In the prototypical Piagetian experiment, the four- or five-year-old will tend to believe that a tall, thin beaker contains more water than a short, wide one, even after observing that an equal amount of the liquid was poured into both. Seeing the higher level of the water in the thin beaker is simply too powerful for him to deny. Thought processes involving the "logical" fact that the amount of liquid in both beakers is the same, since an equal amount was poured into each, are too abstract to be sufficiently compelling.

Slowly, in later childhood, an understanding of the logic of classes and relations emerges. A child gradually gains an increasing ability to withstand the influence of perceptually prominent but irrelevant features of the objects that surround him. Finally, he comes to organize his experiences into an integrated set of cognitive operations that bring a new degree of unity and consistency to his dealings with the environment. With the consolidation of these "concrete operations" at the age of eleven or twelve, a child no longer gives the impression that he understands something only to reveal two sentences later that he does not understand it at all.

Concrete operations, then, become the means for structuring immediately present reality, for dealing logically and consistently with concrete objects. Although a child at this stage is now readily able to draw inferences from the actual to the potential, "the role of *possibility* is reduced to a simple potential prolongation of the actions or operations applied to the given content" (Inhelder and Piaget, 1958, p. 249). The concretely operational child is able to anticipate further ramifications of present events, to construe out of his experiences the means-ends relationships that hold in his environment, and thus to bring images of future consequences to bear on his present actions and to delay gratification for short and specified periods of time. But the pre-adolescent is not yet prepared to apply this same logic to verbal propositions devoid of concrete content. In Bruner's (1960, p. 37) words, "he is not yet readily able to deal with possibilities not directly before him or not already experienced." A child's images of the very different future that awaits him in adulthood are thus likely to remain detached from his conceptions of the present.

As a result, in forming his images of adult life, a pre-adolescent might well fail to maintain a clear distinction between realistic expectations and conceptions that reflect unrealizable, wish-fulfilling fantasies. In Lewin's (1951, p. 141) words, "the 'ideal goals' and 'real goals' for the distant future are not much distinguished, and this future has more the fluid character of the level of irreality."

In a developmental theory of occupational choice, Ginzberg and his colleagues (1951) presented a similar conception of a child's perspective on the distant future. They suggested that qualitative transformations occur in the way a growing person thinks about his vocational future. The ages six to eleven comprise the period of "fantasy" choices. "Tentative" choices are made by adolescents, and "realistic" choices mark the period of early adulthood. Pre-adolescent children, they suggested,

> have been forced to accept the frustrations and limitations which are theirs by virtue of the fact that they are children. However, they have a possibility of escape: they can fantasy about the future. In fact, the child's approach to his future occupation is largely in terms of assuming in fantasy the role of an adult. By playing the role or imagining himself in adult activities, he can help to compensate for the shortcomings of the present (pp. 62–63).

The picture of a child's "future" that emerges from these considerations is one of a contrast between short-range and distant anticipations. By the close of childhood, the relatively near future is well delineated. A child is likely to discern a clear continuity, for example, between his present experiences and his passage through the age-graded sequences of schooling that await him. His ability to delay gratification for short and specified periods of time is similarly well established. His images of short-range future rewards are now imbued with a sufficient sense of reality to justify the rejection of immediate but lesser alternatives.

A child's adult years, however, are likely to remain beyond his reach, standing isolated and unconnected with his actual experiences in the present. As such, they remain available for the projection of wish-fulfilling fantasies, offering a possibility of escape from the frustrations of current reality. It may well follow that the more unhappy and maladjusted a child is in the present, the more likely he is to be oriented toward fantasies of a more pleasant future.

Few children are likely to consider everything that arises in their imagination to be part of the "future." Compared with a child of four or five, a pre-adolescent has considerable sophistication. His conceptions of "the connections which reality offers" are far more accurate. He is far better able to distinguish those imagined experiences that reflect the influence of fairy tales and daydreams from those that conceivably may occur in the future. Yet without the capacity for dealing logically with abstract if-then propositions, and without the pressures that will later induce him to consider seriously the long-range implications of his present actions and decisions, a child may remain relatively unrealistic in his anticipations of adulthood. Not until adolescence are images of that future likely to become more rigidly defined to include only those "conceivable" events that are reasonably likely to occur.

The Empirical Evidence

Davids and Parenti (1958) collected story-completions from boys aged seven to thirteen. They found that indications of current unhappiness, such as the instability of friendship choices and psychologists' ratings of pessimism or resentment, were positively correlated with the span of time encompassed by the action the children described: The more well-adjusted boys generally told stories in which the action was confined to the present. The data were somewhat inconsistent, but the patterns they revealed are important. They call, in the authors' words, for "a re-examination of the hypothesis that future orientation in fantasies of 11-year-old children is a sign of a mature and healthy adjustment pattern." Indeed, these particular children were the inspiration for the study we shall now describe (Klineberg, 1967).

Two groups of French school children participated in this research. One group consisted of twenty-four boys, aged ten to twelve, who were enrolled in two special schools for maladjusted children (schools that the French call "Instituts Médico-Pédagogiques").

Twenty-three boys of the same age and grade levels who were attending private schools in the Paris suburbs comprised our second group. Compared with the maladjusted boys, the children in this group were generally of higher socio-economic status and tested intelligence. They were also getting better grades in school and were free from any seriously debilitating psychological problems. We wanted to know whether advantages such as these are in fact associated with a more restricted orientation toward the distant future before the attainment of adolescence.

The forty-five boys were individually interviewed with a pre-tested questionnaire with which we sought to measure several different aspects of their conceptions of the future. We showed them two TAT cards and asked them to tell us a story based on each, and to indicate the amount of time that had passed between the beginning and the end of the action they described. We asked them to remember ten different things that they had thought or spoken about during the preceding week, and we questioned them about each event until we could determine unambiguously its temporal focus (past, present or future) at the time they had thought or spoken about it. We asked them to list as many different things as they could that might happen to them during the rest of their lives, and for each event to indicate how old they thought they would be when it occurred. Then we gave them our own list of fourteen common life experiences and asked them to guess when each of these incidents might happen to them. Finally, as already mentioned, we measured the consistency with which they ordered these future events in time and their willingness to delay gratification.

There was no difference between the two groups of boys on these last two measures, which is why we felt justified in combining them into a single group when we sought to examine the relationship between delaying capacity and the sense of reality with which images of the future are imbued. There were also no significant differences between them in the number of events they could think of which might happen to them in the future, nor in how far ahead they looked in listing their anticipations. The differences were significant, however, on the central measures of their orientation toward the more distant future of adulthood.

When asked to guess at what point in their lives those fourteen future events would occur, the more maladjusted children made median estimates that extended significantly farther into the future, and they told stories based on the two TAT cards that encompassed a far longer span of time than did the normal boys. On closer examination of these TAT stories, it appeared that the maladjusted boys were far more likely to tell stories with clearly optimistic endings, whereas the normal children were more often content with a description of the situation depicted on the card, often with no discernible outcome at all. Unlike their more maladjusted peers, these children seemed perfectly willing to leave all the action in the present.

Here, for example, is one of the stories told by Claude, a maladjusted boy aged eleven years eight months, in the fifth grade. The picture we showed him was of a small boy sitting at a desk with a violin lying on it:

> It's a little boy who wants to learn how to play the violin, but he can't find any friends to teach him how. He decides to study by himself so that later he can play in a big orchestra. But learning all alone is difficult, and he decides to change instruments. He takes up the harmonica. Then he learns how to play, and later he goes into a big orchestra. (Time span of the action described: ten years.)

Xavier, one of the normal boys, just one month older than Claude and also in the fifth grade, told the following story:

> It's a little boy whose parents want him to play the violin, but he doesn't want to. The parents are not very happy about this, and they ask the music teacher what they can do to make him practice. The teacher says that if he studies the violin, he could become famous. So the parents tell this to their son, but he

still refuses to become a great violinist. The next day, they return to the teacher to ask for another solution. The teacher tells them to get really tough, to take away his dessert every day and to stop him from watching his favorite television programs. This time, the child really wants to study the violin, because television is one of his favorite things. (Time span of the action described: two days.)

These stories are broadly typical of the differences we found between the two groups of boys. Claude is clearly eager to get the child into adulthood, which he envisions as a simple fulfillment of childhood dreams. Xavier, on the other hand, is content to leave the child in the present, and to conclude on a relatively unhappy note.

In general, our results were consistent with the hypothesis that images of the distant future are available during childhood for projecting wish-fulfilling (therefore optimistic) fantasies and that the likelihood of using them for this purpose is enhanced by maladjustment in the present. Also consistent with this notion were two additional findings we have not yet mentioned. First, within the group of maladjusted children, the boys from higher socio-economic backgrounds told stories covering a significantly shorter span of time. Second, among all forty-seven children, those who were doing badly in school were able to list significantly more events that they thought would happen to them in the future, in comparison with their more successful peers.

When things are going well in the life of a child, the distant future seems to be of comparatively little interest. It offers a welcome haven of relatively unconstrained fantasy, however, for those who suffer more acute frustrations in the present.

THE EXPERIENCE OF ADOLESCENCE

The span of years between twelve and twenty is usually accompanied by a series of far-reaching transformations in intellectual ca-

pacities, in role expectations, and in the emerging sense of self. Together, these new developments spell the dissolution of the naïveté
that once informed the child's anticipations. They point to the emergence of a new perspective on the future as a whole.

Cognitive Acquisitions

Piaget's research indicates that the hard-won equilibrium of concrete operations begins to dissolve at age eleven or twelve, when the
final transition to "formal operations" begins. The awareness a child
acquires at this stage is fundamental. It enables him to deal realistically
and logically with conceptions of the distant future, with images and
possibilities whose only reality is in the words he says to himself.
As Inhelder and Piaget (1958, p. 252) have shown, "the most prominent feature of formal thought is that it no longer deals with objects
directly but with verbal elements."

With the development of formal operations, an adolescent no
longer confines his logical intelligence to the real and the present.
When confronted with a conceptual or empirical problem, he can
generate the full range of alternative explanations that the situation
evokes. He can consider logically and consistently what would follow
from each hypothesis if it were true, and he can then direct his
observations by possibilities that exist only in thought. Flavell (1963,
p. 223) summarized this development as follows: "The child deals
largely with the present, with the here and the now; the adolescent
extends his conceptual range to the hypothetical, the future, and the
spacially remote."

In constructing his personal future, the adolescent is now able for
the first time to organize all of his anticipations in terms of logical
thought and realistic considerations. "The possibilities entertained in
formal thought," Inhelder and Piaget (1958, p. 255) wrote, "are by
no means arbitrary or freed of all control and objectivity. Quite to
the contrary."

The connection indicated by the words "if . . . then" (inferential implication) links a required logical consequence to an assertion whose truth is merely a possibility. This synthesis of deductive necessity and *possibility* characterizes the use of possibility in formal thought, as opposed to possibility-as-an-extension-of-the-actual-situation in concrete thought and to unregulated possibilities in imaginative fictions (pp. 257–258).

The acquisition of formal operations makes it possible for an adolescent to extend the span of future time that he is able to integrate with his present experiences. The years of adulthood as the adolescent now conceives them are viewed with a far greater sense of reality than they were when the "real" future was limited to concrete extensions of the present situation or to events already experienced. Images of that distant future are likely for the first time to be products of "logical" rather than "autistic" thinking (Bleuler, 1912).

There is another, more specific, development that may also be relevant in this connection. The integration of distant future events with past and present experiences is probably made easier by an overarching conception that views time as an abstract continuum, encompassing all its separate moments in an irreversible succession. As Wyatt (1964) observed, "the future is really nothing but a slightly normative fantasy in individuals, tied to a hypothetical time continuum." There are other ways of conceptualizing time, as we shall see in Chapter 6. It seems likely that this hypothetical continuum enhances our tendency to regard the future as linearly and causally connected with present and past experiences. The full acquisition of an abstract notion of time in the life of an individual appears to depend on the attainment of formal operations.

Wallace and Rabin (1960) reviewed a large part of the published research concerned with temporal experience. They reported that "the time concept, with ever-widening past and future references, continues to develop through the 13th or 14th year when the adult concept first emerges. At that time the notion of continuity of time and its

relatively accurate estimation are reached." Michaud (1949) interviewed an extensive sample of French school children, asking them
to explain what happens to time when clocks are moved ahead one
hour in the spring and back again in the fall. It was not until age
fourteen that over half of the children revealed an understanding of
time as an abstraction that expresses the underlying continuity in all
events while remaining totally independent of human control. As
Fraisse (1963, p. 280) observed, only with the acquisition of formal
operations during adolescence "is the child capable of passing from
the concrete homogeneity of the time of the clock to the abstract
homogeneity of a duration which is the thread linking events without
being dependent on them."

An adolescent now conceives of time in abstract terms. The future
is that part of the continuum that is open to reasonable expectation
and planning. With these conceptions, he is more likely to sense the
"threads" that bind his present and future together, and the span of
temporal integration may extend dramatically farther into the distant
future.

Meanwhile, other aspects of experience are also changing with the
advent of adolescence. The child has become a "young man" or a
"young woman," with all the changes in external pressures and role
expectations that this social transformation entails. As we explore next
the impact of the "adolescent" role on conceptions of the personal
future, we shall focus on middle-class Western society and on boys
rather than girls. The influence of differences in culture and social
class is examined in Part 3 of this book. The consequences of sex-
role distinctions are explored briefly later in this chapter and more
fully in Chapter 4.

Role Demands

Whenever adolescence is given any separate recognition at all, it
is invariably defined as a phase of transition, of preparation and
training for adult roles. A young person is likely to find himself

suspended somewhere between a relinquished childhood and an unattainable adulthood. In Western societies in particular, this period has become inordinately long and problematical, a period during which a young man or woman is kept in the shadow of significance and usefulness yet expected somehow to construct a meaningful identity. An adolescent is called on to make a difficult and ever-narrowing series of concrete choices involving education, career, residence, and marriage—all in terms of his anticipations of the future. "The social role of the adolescent," Roger Brown (1965, p. 232) observed, "requires him to deal in possibilities, to entertain alternatives and envision consequences. Formally operational intelligence enables him to do so."

An adolescent is likely to find that his present activities are now suddenly treated by parents and teachers as if they were harbingers of the future. Mussen, Conger, and Kagan (1963, p. 563) suggested that

> The very young upper-class child who wants to be an iceman, or fireman, or policeman, may be indulged or even encouraged. After the attainment of adolescence, however, when the problem of vocational choice becomes a serious one with practical implications, the child's parents are not likely to find such notions amusing.

If formal operations underlie the capacity to deal realistically with images of the distant future, the role demands that accompany adolescence provide a forceful impetus to make use of these abilities.

Parents and teachers, of course, are not the only people who define the role demands of adolescence. From the day an American child first enters school, he spends increasing time and emotional energy interacting with his age-mates. By adolescence, the peer group has become perhaps the prime socialization agent, operating independently and often in opposition to the values and standards of the adult world.

Many observers have painted a picture of the "youth culture" as a haven for seekers of immediate gratification. Much has been written of the impulsiveness and rebellion of youth, their excessive conformity to fads, their drive for immediate "kicks," and so on. To the extent that these stereotypes are at all accurate, they may reflect in part an adolescent urge to escape from adult pressures that call for sober reflection about an often uninviting future. In part also, as Ernest Campbell (1969) suggested, the activities and values of the peer culture have a very solid here-and-now reality about them. They thus may offer "useful protection from the implication that the present is a meaningless and worthless state except as it is used to qualify for some desired, remote condition of the future."

Keniston (1965, p. 405) pointed out, however, that most American adolescents, although eagerly participating in peer-group activities, also

> take for granted that they will one day enter upon adulthood, and see themselves as preparing themselves for it. Many of the controversies over the real nature of American youth—over whether it is irresponsible and hedonistic or sober and dedicated —stem from this double orientation. Some observers see one face of youth, and other observers the other, and both observers often mistake the part for the whole.

The part that is of primary interest to us concerns the manner in which adolescents now construct their futures when they feel impelled to reflect on the adulthood that awaits them. We have pointed to certain aspects of their role that seem likely to force on them a more realistic and sober view than was generally the case when they were children.

One result of these new role demands is that adolescents are likely to be made more aware of the future implications of their social position. Many studies have revealed that children of North American Indian tribes show fairly normal educational achievement until adolescence, at which point they come to realize the absence of realistic opportunities and lose all interest in school (see Vernon, 1969). More generally, Margaret Mead (1956, p. 371) suggested that adolescence brings to many disadvantaged youth a growing realization that the goals proclaimed by the middle-class community are not the wonderful opportunities they are so often represented to be.

> So bright girls strangely have no "ambition" and children of discriminated-against minorities turn "dull" at adolescence, not because of intrinsic incapacity, but because the desire to learn is blocked by the knowledge that part of the pattern to which they aspire will be denied them.

The transition from childhood to adolescence in Western society is marked by the introduction of new role demands that define a set of social pressures operating to link conceptions of the future more securely with present realities. No longer can an adolescent indulge whole-heartedly in the childhood fantasies of earlier years. But more than a change in external pressures is involved in this transformation.

No society can long impose a new set of social definitions on a person without in the process affecting his conception of himself. Adolescence, in Erikson's (1959) view, is the stage when the formation of a "sense of ego-identity" becomes critically important, when the effort to develop a coherent sense of self is "precipitated both by the individual's readiness and by society's pressure."

Self-Conceptions

Changed conceptions of self are implicit in the new role demands that confront adolescents. Their role forces them to make decisions now in the light of their conceptions of the future. An adolescent, as Campbell (1969) pointed out, "must make choices, know that he is making them, and know that some of them may be irreversible." Charlotte Bühler (1968) observed that adolescence is the stage when "there is a first grasp of the idea that life represents a time unit with a beginning and an end. For the first time, the youth thinks of his life as something that belongs to him." As Allport (1961, p. 126) suggested:

> The core of the identity problem for the adolescent is the selecting of an occupation or other life goal. The future, he knows, must follow a *plan,* and in this respect his sense of self-hood takes on a dimension entirely lacking in childhood.

Whatever else the concept of "identity" may entail—and it is as difficult to define as it is suggestive in its implications—, it refers in part to the formation of a stable and coherent sense of self within the continuum of time. It involves an integration, as Douvan and Adelson (1966, p. 19) expressed it, of "both the past-in-present ('What I am through what my parents are') and the leap from present to future ('What I deeply hope to be, what I deeply dread being')." In Erikson's (1964a, p. 91) words, "the young person, in order to experience wholeness, must feel a progressive continuity between that which he has come to be during the long years of childhood and that which he promises to become in the anticipated future."

Three central aspects of the adolescent experience thus appear to converge in a single prediction. The attainment of adolescence brings important transformations in cognitive capacities, in role expectations, and in self-conceptions, each differentially emphasized by writers of different theoretical interests and persuasions. Together, they point compellingly to the notion that an adolescent's orientation toward the distant future will be substantially different from that of a child.

THE GROWTH OF TEMPORAL INTEGRATION

In adolescence, as MacIver (1962, p. 36) noted, "the present takes on a definite responsibility for the future." Images of future experiences are now more securely linked with the present, and the adolescent must come to terms with his dreams. Where wish-fulfilling fantasies once reigned serenely, reality now intervenes. Spiegel (1958) summarized the psychoanalytic theory of adolescent development: "What happens to the ego in the years from twelve to twenty? In general, the growth of the reality principle and the triumph of the secondary process." Stone and Church (1957, p. 319) put it this way:

> All his life the adolescent has been hearing the question, "What are you going to *be* when you grow up?" Now the question has suddenly become, "What are you going to *do* when you finish school?" The sense of panic that this question arouses may be enough to send his ideals and ambitions toppling.

Kurt Lewin (1951) was one of the earliest behavioral scientists to emphasize the importance of time perspectives in personality development. He was particularly impressed by the transformations that occur with the attainment of adolescence.

> In adolescence a definite differentiation in regard to the time perspective is likely to occur. Within those parts of the life space which represent the future, levels of reality and irreality are gradually being distinguished. That which is dreamed of or wished for (level of irreality in the future) becomes separated from what is expected (level of reality in the future). Vague ideas have to be replaced by more or less definite decisions in regard to preparation for future occupation. In other words, one has to "plan": to structure the time perspective in a way which is in line both with one's own ideal goals or values and with those realities which must be taken into account for a realistic structuring of the plane of expectation (p. 141).

There is therefore a broad theoretical basis for believing that, as a person grows from childhood into adolescence, a far-reaching change will take place in the way he envisions his future. Adolescence brings important transformations in three cental aspects of experience. The acquisition of formal operations provides a growing person with a new mastery of possibility and brings the verbal elements that express his images of the future under the constraint of logical thought. New role expectations press him to link his conceptions of the future more securely with the implications of his present experience. The formation of a "sense of ego-identy" entails an awareness for the first time of one's entire life span and the effort to bridge the past and the future through a coherent sense of self-continuity. The joint outcome of these developments is likely to be a dramatic extension of the span of temporal integration.

The Present in the Future

If we are right in this conclusion, images of adult life will have acquired, by the age of fourteen or fifteen, a new sense of reality. Anticipations of the future will now be largely restricted to realizable objectives and realistic expectations. Increasingly closed to wish-fulfilling fantasies, they will derive instead from the realities of the present, reflecting the long-range implications of current experiences. Only now will past failures or present feelings of pessimism fore-shorten perspectives on the future and induce a narrower and more defensive present orientation.

Empirical support for a part of these expectations comes from Lehman and Witty (1931) in a study of the vocational preferences expressed by some 25,000 boys and girls, aged eight to eighteen. The authors reported a striking similarity between the developmental curves describing the onset of puberty and those showing a disinclination to enter what may probably be regarded as "fantasy" occupations (e.g., movie actress, cowboy, circus performer, sheriff). There was a sharp and sudden drop in preferences for occupations such as these shortly after age ten and a half for girls and eleven and a half for boys. In the authors' words, "One fundamental fact disclosed by the present study is that children's vocational attitudes mature relatively rapidly after the onset of pubescence."

More recently, Lessing (1968) analyzed the different kinds of events that school children in grades five, eight, and eleven said would occur in their futures. She found a significant decrease, with age level, in the amount of "wish-fulfilling fantasy" the children expressed in their anticipations. As they grew older, they included increasingly fewer material acquisitions among their anticipations and focused instead on educational or vocational goals.

In an effort to test these hypotheses more completely, we interviewed two groups of French adolescents, aged thirteen to sixteen, in addition to the forty-seven children described earlier (see Klineberg, 1967). One group consisted of twenty-one adolescents attending a private school for "scholastic rehabilitation," which caters to boys of high tested intelligence who, for one reason or another, had experienced a great deal of academic difficulty in the public schools. Twenty-three boys of the same age and grade levels at a nearby Catholic boarding school comprised our second group. Each of these forty-three adolescents was interviewed individually with the same questionnaire we used in talking with the younger boys. Their answers reflected a very different way of anticipating the future.

For one thing, the adolescents as a group made significantly more moderate estimates of when each of the fourteen different future events would occur in their lives. In addition, they were far more consistent in their ordering of these events. Both of these measures, as we noted earlier, appear to reflect the degree to which such future events are endowed with a sense of reality.

We also asked all ninety boys to list ten different things they had thought or spoken about during the preceding week and to describe them in sufficient detail so that we could determine whether these were past, present, or future events. It seemed to us that a predominance of present events in this list might reflect a relatively contented involvement in ongoing experiences. If so, we could make contrasting predictions for the two groups of respondents.

Before adolescence, a concern with distant future events is primarily motivated by the desire to "escape" from the present. Among the younger boys, therefore, the more involved the child is in his present experiences, the less likely he is to be oriented toward the distant future. With the attainment of adolescence, however, conceptions of the personal future derive from present experiences and reflect the implications for the future that emerge from that reality. It should follow that the more involved the adolescent is in his present, the more likely he is to be oriented as well toward relatively distant future goals. An unhappy adolescence should lead to a relative avoidance of thoughts and conversations concerning both present experiences and their long-range implications, whereas similar difficulties in childhood should lead the younger boys to envision optimistic possibilities in a future that is still available for the projection of wish-fulfilling fantasies.

We had four different measures of our respondents' orientation toward events in the distant future. We had asked them to list all the things they could think of that might happen in their lives and to indicate how old they would be when each occurred, so we knew how far into the future they stretched their imaginations. We had also asked them to guess how old they would be when each of fourteen different events would occur, and this gave us another measure of prospective "extension." Finally, we had their estimates of the span of time encompassed by the action they described in their two TAT stories. All that remained, then, was to test the relationship between each of these four measures of future orientation and the proportion of present references included in our respondents' reports of ten recent thoughts or conversations.

The correlations, we found, were consistently negative among the children in our sample: The more involved the younger boys were in their present experiences, the less their orientation toward distant future events. Among the adolescents, however, these same correlations were now consistently and strongly positive. It does indeed look as if present experiences and conceptions of the distant future are closely linked by adolescence in a way that is clearly not the case among younger children.

Another set of predictions we made concerned differences between the two groups of adolescents. Here, too, we expected the opposite of what we had found among the younger boys. By middle adolescence, the more unhappy and maladjusted a young person is, the less oriented he is likely to be toward the distant future. Images of that future now derive from the implications of present experiences, and an unpropitious present should lead to a relative avoidance of distant anticipations.

In comparison with the normal young men of the same age, the maladjusted adolescents were able to list significantly fewer events that they thought might happen to them in the future. In addition, when asked how old they thought they would be when the same fourteen events occurred in their lives, these adolescents made estimates that were significantly less distant. By middle adolescence, moreover, they told stories in response to both TAT cards that encompassed a significantly shorter span of time. The stories with optimistic endings and long prospective time spans were now told by the normal adolescents, whereas the maladjusted boys, aged fourteen and fifteen, generally told pessimistic stories within a relatively narrow frame of time.

Here, for example, is one of the stories told by Daniel, a maladjusted adolescent, aged thirteen years ten months, in the ninth grade at the school for "scholastic rehabilitation." The picture was once again of the small boy with a violin:

> It's a young boy whose ambition is to be a businessman. But he is pushed by his grandmother, who adopted him, into becoming a violinist. One day, while he was sitting in front of a sheet of music, the idea came to him of escaping from his home. Leaving his music behind, he ran away empty-handed. His grandmother, when she learned of his flight, simply couldn't understand it. Distressed and frightened, she begged him to return, promising him anything he wanted. But the boy, knowing his grandmother's character, didn't want to go back. Three days later, while he was living in poverty as a shepherd, he

thought about what he had done. Understanding how wrong he was, he decided to return. While he was on his way home, he learned that his grandmother had died of sorrow from having lost him. Frightened by this news, he didn't want to return for fear of being punished. Five days passed with no trace of him. But one evening, his little body was discovered, drowned in a stream. (Time span of the action described: eight days.)

The stories told by the normal adolescents, on the other hand, were strikingly reminiscent of those we received from the maladjusted children. This one, from a boy at the Catholic boarding school, just one month younger than Daniel and also in the ninth grade, is typical:

It's a boy who's wondering what a violin is. He is trying to pierce the mystery of the instrument and wondering if he'll have to learn how to play it. A music teacher arrives, introduces him to the violin, and tells him that he will learn how to use it. He touches the strings, and harmonious sounds arise. The next session comes, and at the end of the lesson the teacher tells the boy that he is very gifted and will have his debut on the stage within a year or two. The great day arrives. The boy is very nervous. Suddenly, one of the back-drops falls down, and a string on the violin breaks. Panic seizes the little boy. He doesn't know if he'll be able to play. Someone brings him a new string, and our friend appears on stage. The audience is angry at the delay, but their mood changes as they listen to the music. He is asked to play again. Success! (Time span of the action described: one and a half years.)

Findings consistent with these were reported by Vincent and Tyler (1965) in a study of ninth graders. Adolescents who listed more distant events in describing their anticipations of the future also scored higher on intelligence tests, gave evidence of greater creativity and a more extensive knowledge of distant places, and participated more in extra-curricular activities. Additional supporting evidence comes from an extensive survey of American high school students conducted by Douvan and Adelson (1966). The authors reported that a major factor differentiating well-adjusted adolescent boys from those who were experiencing more serious difficulties was the ability to bring anticipations of future events into the present and to integrate the past, present, and future into a single cognitive framework.

The available data are consistent with the notion that images of the distant future assume a new reality in adolescence. When self-evaluations are negative or realistic prospects for the future are bleak, the span of future time that may be integrated into present experiences will be affected. An unhappy adolescent, aged fourteen or fifteen, is now likely to restrict his perspective to the surer rewards available in his present surroundings in an effort to avoid the discomfort generated by anticipations of an unpropitious future.

The Lingering Fantasies of Adolescent Women

The transformations in the way an adolescent constructs his future are attributable in part to the new demands of his social role. He is called on to make a series of difficult and often irreversible decisions on the basis of his anticipations of the future, anticipations that thus become increasingly linked with his present experiences. What would happen, then, if adolescents were somehow to find themselves confronted with role expectations for which their present experiences were largely irrelevant? Would we find that images of the distant future will then remain unintegrated with current realities and thereby continue to be available for the projection of wish-fulfilling fantasies?

Until very recently, American society provided precisely the conditions for a "natural" experiment to answer this question. Although traditional sex-role distinctions are today being vigorously challenged, it was not very long ago that a young woman's aspirations were expected to center exclusively on visions of life as a devoted wife and mother—goals that cannot be elaborately planned and for which present efforts have only tangential relevance.

In their national sample survey of over 3000 American adolescents conducted in 1955 and 1956, Douvan and Adelson (1966) discovered striking sex differences in the reality implied by anticipations of the future:

> A boy must have a clear picture of how he will get to his goals; he must reach them, by and large, through his own skills and talents, by his own industry. This fact acts in some way as a check on the boy's tendency to dream. The girl, on the other hand, will reach her goal primarily through marriage. . . . The next steps for her have less to do with personal achievement than with being chosen as a mate by an appropriate young man (pp. 45–46).

If this is true, then the span of integrated time must be psychologically foreshortened for these young women. Only images of that part of the future that precedes marriage can be meaningfully connected with their present activities and current strivings. They can rarely plan realistically for marriage and beyond. Indeed, Brim and Forer (1956) found in their analysis of a still earlier national sample of high school students that girls reported having planned their lives to a significantly shorter distance into the future than had boys. More recently, Lessing (1968) asked her respondents to list ten events they expected would occur in their lives. The boys looked significantly farther into the future in listing their anticipations than did the girls.

Beyond this more limited span of temporal integration will lie a future structured only in vague and tentative ways. Ezekiel (1968) collected three fictional autobiographies from forty-nine Peace Corps volunteers just before overseas assignment. They were asked to describe their plans over the next five years, their life after Peace Corps service focusing on five years from now, and their life at age forty. The essays were scored for "differentiation" (the complexity and detail included in their portraits of the future), for "demand" (the degree to which the future was seen as demanding a long-term and continuing response to challenge), and for "agency" (depicting oneself as the primary agent of the major decisions to be made in the future). On these last two dimensions, the women in the sample scored significantly lower than the men in all three of their autobiographies, but they showed less "differentiation" only with regard to their third essay. There were no differences in their ability to discuss in detail and complexity either former plans or plans for the immediate future. It was in their scores on essays describing their lives at age forty that the women differed most from the men.

When images of future events maintain only a tenuous connection with present realities, moreover, they remain available as vehicles for wish-fulfilling fantasies. Douvan and Adelson (1966) reported that the girls in their extensive sample showed far less realism than the boys when all were asked to imagine a future life for themselves—with the significant exception of those few young women who remained committed to long-range occupational goals. American girls in 1956 generally tended to indulge in stereotyped and romantic visions of suburban bliss, whereas boys were being forced by their social role to view their future largely in terms of vocational goals linked with the realities of their present experience.

The results of this "natural" experiment appear to underline the importance of role demands in effecting the transformations that occur among boys in their conceptions of the future as they grow into adolescence. They also raise the issue, as yet largely unexplored, of the degree to which modern testing and anxious parents can impose a more sober and realistic view of the future on children at increasingly early ages, before the cognitive and social developments usually associated with adolescence have occurred. Teenage girls are clearly not retarded in their acquisition of the cognitive abilities necessary for logical and realistic conceptions of future possibilities. They have differed from boys primarily in the fact that they have been less consistently compelled by the demands of their social role to make a careful and considered examination of their future prospects. Indeed, they have usually been forced to abandon vocational aspirations and

encouraged instead to indulge in romantic fantasies of future fulfillment at the very time when boys find themselves pressed to make
difficult decisions on the basis of their expectations of the future. The
contrasting degree to which a clear sense of realism characterizes their
images of the distant future is a consequence of these sex-role distinctions. As such distinctions fade, one would expect sex differences in
adolescent conceptions of the future to fade as well.

Still, these are primarily differences in timing only. Wish-fulfilling
fantasies may well linger somewhat longer in the images a young
woman constructs of her personal future, but as adolescence gives
way to adulthood such sex differences will largely disappear. The
process of bringing one's aspirations for the future into line with
reality is a continuous one that proceeds, for men as well as women,
well into the adult years. As we have seen in this chapter, it is a process
that typically does not begin until the attainment of adolescence.

There are other, more subtle and more lasting temporal qualities
that are inherent in the social roles through which American men
and women generally come to define themselves. These qualities are
explored in the next chapter.

4
Temporal Orientation, Integration, and Imaginings: The Influence of Sex-Role Learning

> One can add that man's Ultimate has too often been visualized as an infinity which begins where the male conquest of outer spaces ends, and a domain where an "even more" omnipotent and omniscient Being must be submissively acknowledged. The Ultimate, however, may well be found also to reside in the Immediate, which has so largely been the domain of women and of the inward mind.
>
> Erik H. Erikson (*Identity: Youth and Crisis*, pages 293–294)

INTRODUCTION AND BACKGROUND FOR THE RESEARCH

We have just seen how the passage from childhood to adolescence involves major transformations in perceptions of time. Perhaps the most important of these transformations is represented by the increasing degree of realism, the gradual understanding of the relationships among past experiences, present activities and opportunities, and future possibilities. Time perceptions change as a function of cognitive developments, but they change, too, in response to social roles—of what it means, in other words, to be a young man or a young woman in one's society (Mischel, 1970).

Generally, men and women in almost all societies—and certainly in ours—are socialized differently, with the result that sex-role distinctions in perceptions and attitudes develop (Erikson, 1964b, 1968; Bakan, 1966; Kagan, 1964; Terman and Miles, 1936). Consider, for example, the prevalent pattern in our culture that decrees that men are "meant" to pursue occupations or careers and to call these pursuits work (see Hartley, 1959), whereas women are "meant" to care for households and raise children and to call these activities "homemaking" rather than work. If a society or a subgroup within it is to perpetuate such distinctive patterns, men must be trained to prepare actively for the future, whereas women must be oriented instead to deal with day-by-day, if not moment-by-moment, tasks (Parsons and Bales, 1955).

Furthermore, in our country, to be called "successful," a man must transcend his childhood background and "make it on his own," as the expression goes. Women, however, are "not supposed" to achieve in this manner. Instead, their role calls for establishing a continuity with the past and future, essentially through childrearing (cf. Slater, 1970; Mead, 1949). From a temporal point of view, a woman's taking her husband's last name when she marries implies that in a way she must forfeit some of her own personal evolution to enhance the continuity of her husband's family evolution. This, too, is a part of a woman's role, and a part affecting her perceptions of time. All the while, young children are watching closely the actions of adults and deriving from their observations and identifications a sense of how they are "supposed" to define their present and future roles, and hence how they in turn will come to formulate their own perceptions of time (Erikson, 1950; Maccoby, 1966). Whether or not men often dream of having the "easy" life of women, or women of sharing the "exciting" achieving life of men, will not be answered in this chapter, although some of the evidence to be presented here suggests that for members of one sex the role responsibilities and accompanying perceptions of time of the other sex may be especially congenial (Tresemer and Pleck, 1972; Pleck, 1973).

The research presented in this chapter focuses on the time perspectives of American men and women aged eighteen to twenty-two. For the most part, these people came from "middle-class" families. All were high school graduates who, at the time of our study, had completed little if any college work (Cottle, 1969a). In addition, only a few were married. Our respondents were corpsmen and corpswomen chosen from a naval training station where they were undergoing a fourteen-week medical education program. Each week in this continuing program, a new group of about thirty-five trainees is brought in, thereby making it possible to test each person in the group. In our case, testing was done during the second week in each group's program. The important point about our sample was that it was chosen with the intention of minimizing the impact of sociological characteristics, so that if sex role differences later emerged in the results we would be more certain that occupation, age, and level of education could not account for these differences.

The inventories we used were administered in group paper and pencil testing sessions with everyone working on his or her own. All the respondents were advised, moreover, that the results of their work would be held in confidence. Over 500 people filled out the inventories and questionnaires. Afterward, they were interviewed individually and offered the chance to learn more about the inventories as well as talk about the purposes of the study.

Some might criticize this sample population on the grounds that it might reflect characteristically masculine attitudes and perceptions. The military, after all, includes a percentage of men far disproportionate to the male percentage in the society at large. Yet it is precisely for this reason that the population was valuable for our investigation. If, with these sociological factors held relatively constant, and if, with the "masculine" world of the military life imposed on both men and women, sex differences in time perceptions nonetheless emerged, then we would have support for the notion that early childhood and adolescent sex-role socialization contributes to adult perceptions of time. To further demonstrate, however, that immediate situational factors that might make women's perceptions resemble men's perceptions had not influenced this earlier training, we compared the military trainees with a sample of civilians similar in age and family backgrounds. Results of this comparison will be reported later in the chapter.

Based on the discussion in the preceding chapters, we can derive three major temporal dimensions of particular relevance to these concerns. First is what we will call a *temporal orientation*. Is it possible, for example, to define certain groups of people as being consistently more involved in or oriented toward one time zone rather than another? (Fraisse, 1963; Kluckhohn and Strodtbeck, 1961.) Are some individuals more concerned with planning their futures than with preserving past traditions, whereas others reveal the opposite tendencies? Are women, perhaps, more oriented to the present, and men generally more concerned with the future?

A second dimension involves the linking of anticipations with conceptions of past and present experiences, or more simply, the degree of connection among the time zones. Whenever we consider such central notions as the sense of personal control over the future—concern with providing for an uncertain future and planning for and trying to anticipate that future—we are in effect saying that the past, present, and future are integrated in different ways and to varying degrees (Murray, 1959). Conversely, when we speak about the failure of imagination to structure realistic images of the future, an inability, that is, to recognize connections between present effort and future outcomes, we are suggesting a lack of integration between the present and future. *Temporal integration,* therefore, became the second major concern of the research.

A third dimension relates to *the imagined future* itself (Ezekiel, 1968). Regardless of how any one person might construct the future, there remain characteristic ways of dealing with its inherent uncertainty. Among these are, first, degrees of *expectation* or planning; second, *fantasizing direct preknowledge* despite the totally unrealistic nature of this action; and third, the willingness to *predict* future outcomes of both a personal and societal nature. Although both expectation and prediction connote presently held responses to the future, the former implies a belief in some future event's occurrence, whereas the latter implies a virtual certainty that it will occur. In predicting, one says, "I *know* I will live to be eighty years old." Preknowledge, on the other hand, is a wishful fantasy, for the best one can say is, "I *wish* I knew how something will turn out."

A fourth dimension is suggested by the investigations of Cohen (1966), Farnham-Diggory (1966), and Wohlford (1964), in which attitudes toward the different time periods are inferred from the way people perceive their duration. If sex-role socialization influences such fundamental perceptions as temporal orientation, integration, and imagination, then it might also affect estimates of the duration of time periods. In other words, how long a particular time period seems to last might also be a function of one's sex-role training.

MEASURING TEMPORAL ORIENTATION

The study began with the construction of an inventory that would enable people to reveal their orientations to the past, present, or future. When instructed to list the important experiences of their lives, do they think of past, present, or future experiences, or all three in some logical proportion? The actual instructions of the Experiential Inventory (Cottle, 1968b) began as follows:

> Please list the ten most important experiences of your life. These may be experiences you have had, you are having, and experiences you expect to have. You only need to write a few words for each experience, and you may list your experiences in any order you wish.

One can see already the problems that arise in developing such an inventory. At almost every point, methodological decisions must be made, and as they are, certain options disappear, eventually cutting into the richness of such an inquiry. Ten experiences, for example, is an arbitrary figure, meant to provide a sufficient number without unduly taxing the person responding. Then, too, the word "expect," in contrast to words like "wish for" or "hope," is problematic, since not everyone will interpret expectation to mean certainty of occurrence.

Imperfect as it was, the inventory was designed to study how a person placed his ten "most important" experiences in time. A second set of instructions was administered after the ten experiences had been listed:

Now that you have listed ten experiences, please study the time zones shown below:

Time Zones

1. Distant past
2. Near past
3. Present
4. Near future
5. Distant future

Now, take each experience and decide if it has occurred, is occurring, or will occur. Then choose the number from the time zone list that best represents the time of the experience and write this number in front of the experience. Do this for all ten experiences.

Again, a number of mechanical and conceptual problems arise. Some people, for example, have been thinking only of the experiences themselves and are surprised by the request to locate them in different time periods. Everyone, however, confronts an ambiguity in the instructions, for how is one supposed to define the time periods? When, for example, does the present conclude and the near future begin? When does the distant past end and the near past begin?

To learn how people judge the duration of time periods, the Duration Inventory was constructed (Cottle, 1968a). The task here was to define in chronological units the boundaries of the five time periods presented in the Experiential Inventory. The units to be employed in these definitions were seconds, minutes, hours, days, weeks, months, and years. Respondents were instructed to use these units to indicate the "duration" of each time period. Keeping this in mind, let us recall that the "present" requires two units, one for its duration backward toward the near past, the other for its duration forward toward the near future. Someone, therefore, may report that the present runs from years ago to seconds from now.

Although the Duration Inventory may seem simple at first glance, several intriguing results emerged from it. First, the present's past and future borders were, for the most part, measured by both men and women either in terms of seconds, the shortest chronological unit, or years, the longest unit, but rarely in terms of weeks or months. Second, distance, as regards both the past and future, was measured almost always in years, whereas units of nearness ranged all the way from seconds to years, although months were the most popular unit. Third, the boundaries of the present were not always equal or symmetrical: How far back in time the present was seen as extending did not necessarily coincide with how far forward it seemed to extend. Specifically, although the prevalent image was of a symmetrical present (that is, the present's duration backward in time was equal to its duration forward), 40 per cent of the men and women described an asymmetrical present. Typically, it was a present with its future border longer than its past border.

One last item about the Duration Inventory seems worth mentioning, as it anticipates an aspect of time perception that will concern us later in this chapter. We discovered that, in filling out the Duration

Inventory, some people believed that a certain mathematical consistency ought to underlie their calculations. If the near future, for example, ends in a matter of hours, they contended, then the distant future should begin in a matter of hours, or some unit longer than hours. Occasionally, however, the units selected revealed a lack of connection among time periods. A present, for example, was seen as ending seconds from now, but the near future did not begin until months from now. A space, in other words, existed between time periods. Other people, however, envisioned one time zone commencing before a contiguous zone had concluded, in a pattern we called overlapping (Cottle and Howard, 1969). In the end, all possible designs and perceptions occurred. Periods were seen as overlapping, as continuous, and as discontinuous. Their duration, furthermore, was judged to be short, long, and anywhere in between.

Despite these intriguing results, our interest lies in temporal orientation and the temporal location of those ten important life experiences. Our procedure for examining the results of the Experiential Inventory involved dividing our respondents into three groups (Cottle, 1969a). The first consisted of those who had listed three or more experiences in the future. We made no distinction here between experiences located in the near or distant future. For want of a better label, we called these respondents "future oriented." This term, however, should not conceal the fact that members of this group had also located many of their important experiences in the past and present as well. The second or "middle" group consisted of people who listed only one or two future experiences. Finally, our major interest focused on those persons who reported no future experiences at all. For whatever reasons, this last group of young men and women listed *only* past and present experiences. Accordingly, they were called "past-present oriented." Presumably, these were people for whom an event cannot be important unless it has already been experienced or is presently being experienced. An expectation or a prospect appears to have too little sense of "reality" to be significant. As it turned out, each group represented roughly one-third of the sample, and each contained an almost equal number of women and men.[1]

MEASURING TEMPORAL INTEGRATION

The second dimension to be measured in the study was temporal integration, or the perceived connection among time periods (Cottle, 1967). Although the Duration Inventory touches on aspects of temporal integration, a different instrument was designed to measure this dimension directly. It began with the following instructions:

> Think of the past, present, and future as being in the shape of circles. Now arrange these circles in any way you want that best shows how you feel about the relationship of the past, the present, and the future. You may use different size circles. When you have finished, label each circle to show which one is the past, which one is the present, and which one is the future.

Our interest was in our respondents' perceptions of relatedness among time periods, but the suggestion that they use varying circle sizes came to have additional meaning. Size, as we confirmed in interviewing the respondents, was equated with perceived importance: The larger the circle, the more important the time zone was felt to be. Relative circle size, furthermore, provided a clue to a person's

[1] Another way to look at the results of the Experiential Inventory would be to compute a mean score by adding the numbers corresponding to the time zones listed for each experience (1 for distant past, 2 for near past, etc.) and dividing the total score by ten, the number of experiences. In this procedure, the higher the mean score, the more future-oriented the experiences. The problem of working with means, however, is that they mask important differences in ways of responding to the Inventory. A score of 3, for example, might indicate that two experiences have been deposited in each time zone, or that five experiences have been placed in the distant past and five in the distant future. We would, therefore, learn little about a person's overall sense of temporal orientation.

sense of the emergence of time periods. For example, time periods could increase or decrease as one went from past through present to future. Or, circles could be drawn in ways that precluded any sense of a gradual enlargement or diminution of time periods. For that matter, circles could be drawn vertically, horizontally, or diagonally, and presumably each of these dimensions also reveals something about a person's perceptions of the past, present, and future.

Given these many avenues of possible exploration, three variables were ultimately derived from the Circles Test. They were "time zone relatedness"; relative size of circles, or what was called "dominance"; and "temporal development," a term describing circles gradually becoming larger or smaller as one moved from past through present to future. Temporal development could be either "future dominant," in which case the future was the largest circle and the past was the smallest, or "past dominant," in which case the past was the largest circle and the future was the smallest. But this meant that we needed confirmation that people had purposely drawn one circle larger than another. When, after interviewing almost 90 per cent of the respondents, we learned that they had indeed purposely drawn circles of different sizes, we awarded two points to a circle for each circle that it exceeded in size. The largest circle, therefore, received four points, the middle-sized circle two points, and the smallest circle no points. In instances in which two circles were the same size, we awarded no points to either.

Relatedness among circles was a bit more complicated to score. The patterns describing the extremes of relatedness were circles not touching at all, a design we called "atomistic," and circles drawn totally within one another, a design we called "projected." In between these extremes lay two possible configurations. Circles could touch at their peripheries, forming what was called a "continuous design," or they could overlap, thereby sharing a certain amount of space. This last design was labeled "integrated." To simplify matters, no distinction was made among degrees of overlapping. The accompanying figure may help in visualizing the various patterns.

Temporal Relatedness Designs from Circles Test

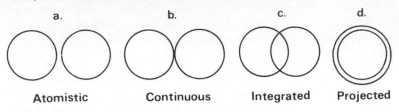

a.	b.	c.	d.
Atomistic	Continuous	Integrated	Projected

We decided to award no points to atomistic circles, two points to continuous circles, four points to integrated circles, and six points to projected circles. Since there was no assurance, however, that a person would draw the same circle configuration across the three time zones, each circle had to be scored individually. That is, the past and present might be continuous (two points), whereas the present and future might be integrated (four points).

Once analyzed, the results of the Circles Test surprised us. Assuming that size would reflect number of years, we had anticipated that given their age, women and men alike would draw the future as the largest circle, which they did, and the past as the second largest, which they did not. We also expected that most drawings would show continuous if not integrated time periods, suggesting some sort of connection among the zones. In fact, over 60 per cent of the drawings were atomistic.

The first evidence of sex differences appeared when men drew circles indicating temporal development more frequently than women. Only among men, moreover, was temporal development significantly associated with both future dominance and integrated time periods. The men who saw time periods increasing in size as one moved toward the future also tended to relate the three time zones and to attribute minimal importance to the past. On the other hand, the women who related the time periods were less likely to view the future as the dominant zone. Notice here that when we say "related the time periods" we mean drew continuous, integrated, or projected designs.

MEASURING PREKNOWLEDGE AND
THE WILLINGNESS TO PREDICT

Because it includes expectations and plans, The Experiential Inventory had already provided us with some indication of how the future was managed. Two additional kinds of future imaginings, however, were explored in the study. First was the fantasy of preknowledge, as well as the fantasy of recovering time from the past. Second, was the willingness to predict future outcomes of both a personal and societal nature.

The various forms of future imaginings that we are considering here represent different approaches to the future. They are suggestive of but a few of the ways in which people handle the future's uncertainties. The fantasies of foreseeing the future and of recovering past time were explored in the Money Game (Cottle, 1969b). Nothing in the instructions relating to this procedure masked our intention of playing a game with time:

> Pretend that you had a lot of money, more money than you could possibly use. Pretend also that someone had the power to sell you time, any time that you would want, and you would have this time given to you right now, knowing what you now know. Furthermore, having this time, you could do whatever you wanted with it.

We gave our respondents the choice of the following amounts of money: $10,000., $1000., $100., $10., $0. These were the prices they would offer in exchange for units of time from four temporal zones. We assumed that the more they wished to recover or foresee a particular period, the more money they would be willing to spend for it.

The units of time consisted of an hour, a day, and a year. The temporal zones included the historical past, defined as the time prior to one's birth; the personal past; the personal future; and the historical future, defined as the time following one's death. Time allotments were used because we felt that some persons might like an hour's peek at the past or future, whereas others might feel that pretending would only be worthwhile if a sufficient amount of time, like a year, were made available. This differentiation in time allotments, however, turned out to be unnecessary.

Two final points about the Money Game should be mentioned. First, the issue of reality was purposely circumvented. Nothing in the instructions suggested that foreseeing the future or recovering time from one's personal past or from the historical past would in any way upset the normal eventualities of one's life. Time, then, could be retrieved as well as foreknown. Second, the instructions made it clear that people could indulge in the fantasies as much as they liked.

The results of the Money Game also contained some surprises. Despite the premise of limitless wealth, as many as one-third of our respondents refused to bid anything for the chance to recover their past or foresee their future. In addition, men played the game more than women, but only with regard to the personal past and personal future. They did not pay more than women for the historical past and historical future, time periods which barely tantalized our respondents at all.

Another result of the Money Game emerged from our interviews. Almost 80 per cent of the men who were willing to play the game to recover their pasts reported that they would seek out those experiences in their lives that had caused present or imagined future difficulties and "re-do" these experiences in order to improve their futures. Approximately 75 per cent of the women, however, tended to select pleasant past experiences that they simply wished to relive. Only about 6 per cent of the women said that they would use retrieved experiences to reshape the present and future.

Our final instrument dealt with the willingness to predict future events. Here the question of what people would predict was not so significant to us as whether or not they would be willing to predict at all. Our concern, in other words, was with a predisposition to commit oneself with some certainty to future outcomes.

The instrument designed to explore this predisposition, the Future Commitment Scale (Cottle, 1969a), consisted of thirty-one statements. Our respondents were instructed to agree or disagree with these statements, or to mark a third option labeled "can't say." The statements described incidents whose outcomes no one presently could know. Seven of the statements referred to personal issues (e.g., "My children will become greatly successful."), and the remaining twenty-four dealt with societal issues (e.g., "Russia and America will live in peaceful coexistence.").

Although a final analysis of the Future Commitment Scale included an examination of all three possible responses (agree, disagree, and can't say), our main interest was in differentiating our respondents according to the frequency with which they chose the "can't say" option. This we took to be an indication of their reluctance to predict future outcomes. The results of the scale indicated that, in general, men committed themselves to future-oriented statements far more readily than women did, but only when the statements involved societal issues. On the seven personal statements, men and women performed similarly, everyone indicating relatively little willingness to make predictions.

RESULTS OF THE STUDY

We have by now generated quite a list of variables. There are, for example, those respondents labeled "past-present oriented," "middle," and "future oriented" on the basis of the Experiential Inventory; those who show temporal relatedness of any kind on the Circles Test and those who do not; those who are willing to play the fantasy game of recovering or foreseeing time and those who are not; and finally, those who predict future outcomes on the Future Commitment Scale and those who believe that future events are inherently unpredictable. The task before us now is to investigate relationships among these various response categories, still with an eye to the effect of sex-role learning on temporal orientation, integration, and future imaginings.

Most intriguing, the significant relationships among the different inventories hold either for one sex or the other, but never for both. Future-oriented men, for example, indicate some degree of relatedness among the past, present, and future on the Circles Test, whereas future-oriented women do not. Conversely, past-present oriented men draw three unconnected circles (temporal atomicity), whereas past-present oriented women tend to draw related circles. Furthermore, although there is no relationship for either sex between the experiential time orientation categories and the fantasy of recovering the personal past, the past-present oriented women reveal significantly less interest in obtaining foreknowledge of the future compared to future-oriented women. Again, however, past-present oriented men do not mirror this pattern. Indeed, there is no systematic relationship between these two measures among the men. Finally, future-oriented men, but not future-oriented women, score low on the Personal Prediction part of the Future Commitment Scale. It is the past-present oriented men only who appear willing to commit themselves to the predictability of the future, scoring significantly higher on this dimension than the "middle" and "future oriented" groups.

THE EFFECT OF VALUING ACHIEVEMENT

Our review of a number of empirically derived time perception variables is almost completed. One last item from the original study, however, should be presented here because it elucidates these inter-relationships and provides a guideline for interpreting the sex differences that emerged.

When we undertook our study of time perception, many attitudes and values struck us as being potentially relevant. We considered, for example, a task that asks persons to recall experiences from their childhood, or a task asking for a description of their life ten or fifteen years hence. We also considered the concept of achievement. What, after all, is achievement if not a way of handling, in a dramatic sense, properties of time?

Many social science investigations demonstrate characteristic perceptions of time held by those revealing a strong need to achieve (McClelland, 1961; Knapp and Garbutt, 1958, 1965; Winterbottom, 1958; Rosen, 1956). In particular, high achievers reveal a special concern with the future as evidenced by their predisposition to make up stories that occur in the future. (See, for example, Epley and Ricks, 1963.) Equally important, in looking at items that comprise the scale designed by Strodtbeck (1958) to measure the presence and strength of the achievement value, one finds that many of these items are themselves temporal in nature. Here, for example, are three items from the eight included in Strodtbeck's Achievement Value Inventory: "Planning only makes a person unhappy since your plans hardly ever work out anyway"; "When a man is born, the success he's going to have is already in the cards, so he might as well accept it and not fight against it"; and, "Nowadays, with the world conditions the way they are, the wise person lives for today and lets tomorrow take care of itself." More subtle temporal implications exist in other items on the same scale: "Even when teenagers get married, their main loyalty still belongs to their fathers and mothers"; or, "Nothing in life is worth the sacrifice of moving away from your parents." Both of these statements suggest an orientation to a family tradition that one is taught to honor. In effect, all the items that are presumed to measure the achievement value reflect, either explicitly or implicitly, ways of handling the future. According to this one scale, therefore, to value achievement is to subscribe to a certain set of beliefs of which time is a major dimension.

Knowing this, it seemed important to explore directly the relationship between the Achievement Value Inventory and our measures of temporal orientation, integration, and future imaginings. For one thing, the achievement items might be used to clarify the time variables themselves. One might discover, for example, that the willingness to move away from one's parents and start a "new life" is a major reason for separating the past from the present and future on the Circles Test. Equally important, the time perception variables might deepen our understanding of what it means to value achievement. To what extent might achievement require a certain configuration of orientations to the past, present, and future?

To determine the relationships between the Achievement Value Scale and the time perception variables, we administered the scale to our respondents some twelve weeks after they had completed the time perception measures. We gave no indication of the possible connections between these two sets of inventories until the Achievement Value Scale had been completed.[2]

The results involving the relationships between performance on the Achievement Value Scale and on the time perception variables are easily summarized (Cottle, 1969c). First, women and men obtained almost identical scores on the scale, so neither sex could be said to value achievement more than the other. Second, women and men alike revealed a relationship between valuing achievement and perceiving a relatedness among time periods (Circles Test): The more people valued achievement, the more they related the circles to one another; conversely, the lower the achievement values, the more likely they were to draw atomistic designs.

A third finding was that future-oriented men (as measured by the Experiential Inventory) valued achievement more than other men did—a relationship not found among the women. This result is consistent with those previously cited studies that demonstrated an association among men between a need to achieve and a generalized orientation to the future. Finally, women, but not men, revealed a relation between valuing achievement and wishing for preknowledge of the personal future in the Money Game.

---

[2] As a methodological check, we administered all the time perception inventories again during the last week of the training program in order to determine whether the inventories yielded relatively stable results. It turned out that the stability was very high.

SPECULATIONS ON THE TIME PERCEPTIONS
OF MEN AND WOMEN

Before suggesting interpretations of these findings, one final matter must be settled. For this research, we had purposely selected a group of people who were similar on several important sociological dimensions, and at an age when they were dealing seriously with intentions and long-range plans. To make certain, however, that it was not the military environment that contributed to the perceptions we have just reported, the entire study was replicated on a different population. This time, our respondents were 300 high school seniors in Chicago and Boston, men and women aged seventeen to nineteen from middle-class families.

The results of this second study were almost identical to those of the military trainees. There were slight changes on some of the inventories, but no differences of a statistically significant nature emerged. More important, the relationships among temporal variables, on the one hand, and between temporal variables and the achievement value inventory on the other, were virtually the same as those we found in the earlier study. On this basis, we may conclude that sex-role identity can indeed account for a certain proportion of the systematic variations we find in temporal perceptions, whereas the military environment, presumably, accounts for far less.

Throughout this and other chapters, we have spoken of the prospective orientation of human beings, the significance, that is, of expectation, hope, prediction, foreknowledge, preparation, or anything that goes into carving out the future. No sociological inventories are required to know that each of us handles the concept of the future in various ways and appeals to experiences of the past and present in order to understand the future. And, if we had any doubts that the future was the dominant time zone for the men and women in our study, the results of the Circles Test have settled them.

It is in the relationships among inventory variables, then, that we obtain our best sense of the way these particular people deal with their future. We cannot conclude from these relationships that a "normal" set of temporal perceptions exists, or that either sex approaches time in a more "suitable" or "healthy" fashion. We can say, however, that a past-present orientation on the Experiential Inventory—a failure, in other words, to list even one future experience—seems to be balanced by characteristic behavior on the part of men and women. The past-present oriented men readily made predictions about themselves or their children, whereas past-present oriented women fantasized a preknowledge of the future. In these ways, presumably, the mystery of the future was somehow reduced for these two groups.

More generally, our findings suggest that this particular group of middle-class women and men have been socialized differently, with the result that they tend to approach the concepts of time and achievement in a fashion consonant with their distinct sex-role identifications. Notice, for example, that although achievement is not a predominantly masculine value in our sample, it was associated for men with the *expectation* of important events awaiting them in the future, whereas for women it was associated with a *wish* to foresee the outcomes of personal events. Separation from the past and especially from one's family, which on the Achievement Value Scale is central to the definition of achievement, was not reflected in the temporal relatedness variable on the Circles Test. To the contrary. Valuing achievement was positively associated with temporal relatedness. What these women and men seemed to be saying, therefore, was that although transcending the past or cutting themselves off from dependent relationships emerge as requirements of achievement, these factors do not imply that the past and present have no connection either between themselves or with the future. Indeed, a strong connection among the three time periods characterized the circles drawn by those who valued achievement highly.

Of all the variables studied, the most suggestive, perhaps, are those derived from the Money Game. It was in the way they indulged in temporal fantasies that these women and men differed most from each other. The fantasy of recovering the personal past among men, the fantasy of preknowledge among future-oriented women, and the correlation among women only between valuing achievement and fantasizing preknowledge suggest a consistent sex-role variation. We may speculate that the socialization of these young men into careers and, more comprehensively, into orientations that throughout adolescence demanded preparation and the delay of immediate gratification produced a sense of having the present severed from the past and tied instead to the future. This, in turn, promoted a desire to return to, or at least regain, the past (cf. Dummett, 1967). The desire to retrieve time through a game of pretending surfaced as a wish that presumably realistic orientations to time ought not tolerate. For, as part of their socialization, these men have learned that realistic preparation for the future is *the* dominant activity, the dominant orientation.

The fact of the matter, however, is that the natural flow of life, at least as it is construed in Western culture, teaches women and men alike that it is toward the future that all activity is directed (Bugental, 1967; Berger, 1967). This special push in the form of a future-oriented socialization process might, therefore, seem superfluous if not actually alienating to men. The women in our sample, after all, also drew the future as dominant on the Circles Test, and many women listed future experiences of one sort or another on the Experiential Inventory. Yet, something about the way these women left their childhoods made it less necessary for them to fantasize the recovery of their past. It may be that the future they visualized would be enough of a repetition of, or replacement for, the childhood that they had abandoned to make the recovery of time seem unnecessary. This suggests that despite their navy obligations, these women generally expect to become wives and mothers and thereby to "fill out" the roles they witnessed as children. The men, on the other hand, expect to assume occupational roles that, as adolescents, they were obliged to imagine, since children, for the most part, rarely have much opportunity to observe their fathers at work (cf. Ames, 1946; Piaget, 1955, 1966).

If these notions seem reasonable, we may further suggest that an important aspect of these divergent forms of socialization patterns is that men are taught to tolerate the ambiguities and uncertainties of the future more than women (cf. Allport, 1963). In particular, they must recognize that if they truly value achievement or hold to the idea that major life experiences await them in the future, then they cannot tolerate the idea, even in fantasy, of knowing what the future holds before it arrives. Wishing to know the future's content would appear to violate the "proper" approach to the future taken by men who value achievement. Fantasizing preknowledge does not, however, seem to violate feminine rules of socialization, because expecting, wishing to know what the future contains, and valuing achievement did not conflict for the women in our sample.

Through the processes of socialization and sex-role learning, women in our society generally continue to be taught that careers and future achievement may remain as items on social-psychological inventories but do not constitute realistic possibilities for their own lives. The realities faced by men, for whom careers and future achievement so often mean everything, are not given equal attention by women. Of course the requisite values, attitudes, and perceptions are all there; but for the most part, the sense of genuine possibility is denied to women by the society's norms, styles, and opportunities.

The injustice of sexual inequalities must not mask a corresponding problem experienced by men. To judge from the results and interviews of our study, men appear deeply alienated from their past and possibly from their present as well. The alienation, moreover, derives from learned definitions of achievement and future preparation. In the case of many men, so much time seems to be devoted to preparing the future that the hour-by-hour, day-by-day activity that describes the process of life is ultimately unsatisfying. The need they have acquired to prepare their lives and forego present gratifications, causes many men to downgrade personal involvements in favor of impersonal and demonstrable accomplishment. Women, on the other hand, are socialized to believe that much of life rests on the capacity to nurture people and to sustain interpersonal relationships. The collectivistic or communal orientation of women, therefore, stands in contrast to the individualistic and instrumental orientation of men (Bakan, 1966; Parsons and Bales, 1955). Not surprisingly, the results of such narrow orientations often turn out to be unsatisfactory for both sexes.

Granted, much of our discussion has been speculative. It is unfair, perhaps, to offer such interpretations in light of the sample we have used and the limitations of the instruments and data we have generated. No doubt, women and men possess singular lives, unique childhoods, and irreplaceable and unreplicable experiences. But, as we shall see in Chapter 6, no experience, whether actual or prospective, assumes form or meaning apart from the symbols of a culture and the structures of a society (Gurvitch, 1964). The process of socialization, of becoming a participating member of a society or of a group within it, implies that women and men are taught how to think about and evaluate memories, present experiences, and prospects. And so, the single life, the unique personality, is channeled by cultural values and social norms until finally, through the allocation of sex-based roles, one begins to recognize characteristically "masculine" and "feminine" perceptions and behaviors within a particular age group and social class.

5

Matilda Rutherford

"YOU CAN'T HAVE A FULL LIFE WHEN THE BEGINNING GETS ALL MESSED UP."

No one should think of childhood merely as a stage of development. Each moment of life possesses its own inherent value, its own integrity, as well as its special capacity to prepare a person for later moments. The studies reviewed in Chapters 3 and 4 reveal the complexity of maturation from childhood to adolescence generally, and in particular, the cognitive and social-psychological factors that influence a person's changing conceptions of time. A certain level of maturational development, furthermore, must occur before children can even speak about their perceptions of time, much less make inferences about the future. Although language and thought are the foundations of temporal contemplations, they will be shaped by sociological factors, by the values and rules of behavior that ultimately give form to young people's conceptions of the future, and to themselves as being in that future. The decision about a career, for example, a decision that society in part determines for its members, becomes a major ingredient of one's self-concept, as well as the concept of self that one projects into the future.

Self-conceptions, moreover, are constituents of one's identity, and identity is built on the dimension of time. Identity depends on continuity and evolution, on history, the same collective history that one begins to fully appreciate only during adolescence (Erikson, 1968).

The integrity and continuity of generations, that Marcus Simpson envisioned, form in adolescence. Language, the development of formal operations, the ability to comprehend the meaning of the word "possibility," the capacity to abstract, the appreciation of history, and the transformation from wish-fulfillment to the active carving out of a present and a realizable future are what characterize an adolescent's involvement with time, and what, in large measure, distinguish it from a child's temporal involvements.

If the preceding two chapters have taught us anything, it is that young people learn how to face realistically the meaning of the future and to adjust their perceptions of the past and present in an effort to manage this new future. The learning that they undergo, however, is not purely idiosyncratic—that is, it is not fitted primarily to individual personalities. Indeed, the study reported in Chapter 4 revealed a series of systematic variations in time perception among young men and women. It is not a question of whether men are more or less future oriented than women, or more or less committed to valuing achievement. The important issue is that as children mature, their cognitive capacities to appreciate the concepts of past, present, and future undergo the sorts of changes that were traced in Chapter 3, and these changes are themselves affected by the sex-role standards of their society. The manner in which young men and women encounter the future is therefore related to styles of masculinity and femininity.

One characteristic pattern of such sex-role differences in perceptions of time is the one presented at the close of Chapter 3. We might recall from that discussion how young women's visions of life were only weakly tied to present activities, and how they planned their lives to a significantly shorter distance into the future than men. Adolescent women experience only the barest connection between present circumstances and future events, with the result that the future, and possibly the past as well, remains available for the projection of wish-fulfilling fantasies.

If young women are taught that the future simply arrives in due time and little need be done to prepare for it, then the reality inherent in a vocation or career should counteract this idea. When a vocation is anticipated, the future tends to be carved out more carefully and less romantically, and the past and present are reconceptualized accordingly. There is now something "out there" to be achieved and worked for. The acts of preparing, delaying gratification, and actively establishing expectations thereby become realistic and even necessary enterprises for women, just as they are for men; or at least they would if society genuinely supported the career aspirations of women as it presently supports those of men.

I met Matilda Rutherford several years ago. Our conversations shifted topics practically each time we spoke, but always her visions of childhood and adolescence and the theme of time dominated our meetings. Middy, as she is known to her friends, lives and works in a poor community of Boston. She comes from what many of us would label, in the social science shorthand of the day, a "broken home." Her closeness with her family in part explains her interest in its history. Her involvement with her mother, moreover, coupled with her present career, accounts in part for her images not only of the future, but of the past as well.

Throughout our discussions, Middy Rutherford drew inferences about the future and about those aspects of the past that she either could not recall or, more likely, never knew. Her language slipped easily from the past to the present to the future; her dreams of an ideal future revealed her competence, and her ability to control events of the present. At times these dreams seemed unreal, fantasy compensations, perhaps, for the unhappiness she had known. At other times, they appeared to be the logical outcomes of a particular family and social history. She seemed concerned with preparing herself for change and watching closely for the realistic opportunities of the future that might alter her most private self-conceptions.

One cannot hear her accounts without being reminded of the images of the future described for young women in Chapter 3 by Douvan and Adelson (1966), Brim and Forer (1956), and Lessing (1968). The wish-fulfilling fantasies and their attending romantic qualities are certainly evident in this one woman's account. Yet, the impact on her of poverty, racism, sexism, sex-role learning, the nature of existing opportunities, and the viability of her career also permeates her accounts of time as she experiences it. We might recall, in this connection, Margaret Mead's words, which summarize a major concern of ours in this section: ". . . and children of discriminated against minorities turn 'dull' at adolescence, not because of intrinsic incapacity, but because the desire to learn is blocked by the knowledge that part of the pattern to which they aspire will be denied them."

Dullness may occasionally be the state of a person that we perceive. More likely, however, as Mead suggests, it is the state of historical, social, economic, and political circumstances, and of our inability to touch another person. Often we perceive a dullness that only our relationships have created. Yet, we might also see a society depriving certain of its members of nothing short of a future (see Chapter 6). What we often overlook are the ways in which we ourselves disparage these same members' sense of the past and the present. And so we leave it to *them* to tell *us* how these bits of time in human experience might fit together.

Her name is Middy, for Matilda, and she's what *you* would call a whore." Once these words served as my only description of a nineteen-year-old black woman. One of her boy friends told me about her, advising me that she would speak for herself. Through him I came to know Matilda Cleveland Rutherford, named after the grandmother who reared her. The Cleveland family had moved from rural Virginia to Boston in the late 1920s. Marcus Cleveland, Middy's great uncle, had been a miner. When the lung disease he carried with him from his early twenties prevented him from working any longer in the mines, he and his brother Case decided to move north. Taking their meager savings, they headed for Massachusetts. Having settled in Boston, they sent for their sisters, Matilda and Ruth.

Within a year in her new home, Matilda Rhodes Cleveland, the oldest sister, was married to a man named Ernest Weaver. Two years later their second child, a daughter named Grace was born. The first child, a boy whose name was to have been Clarence Rollins Weaver, was born dead. At the time it was said that proper medical attention would have saved him. Both children were born in the Weaver home, on the floor, actually, of the front hall where the wood was freshly painted. Three more children would be born in that house, all of them in the master bedroom in the rear. Two of them would survive.

Grace Weaver remains a beautiful woman. Tall and slim, she could easily be mistaken for someone in her late twenties, a fact that continues to annoy her more than please her. When she was sixteen, and a sophomore in high school, she became pregnant. A son was born to her despite the protests of members of her family, who urged her to have an abortion, or at the very least to give the baby up for adoption. The father of the child, Armand Jenkins, disappeared shortly after learning of the pregnancy. Once, as a surprise, he sent his son a ten-dollar bill. No address accompanied the gift, however. The postmark read El Paso, Texas.

Matilda and Ernest Weaver helped their daughter care for the baby, a plump and happy boy. There were many discussions then about morality, prudence, and maturity, but throughout the experience the elder Weavers remained willing to treat their grandchild as precisely that.

Despite her parents' admonitions and her pledges to herself, Grace Weaver became pregnant again a little less than two years after the birth of her son. She tried as best she could to conceal the fact from her parents, but after four or five months there could be no more concealment. Furious, her father threatened to disown her. Holding down a good job, he was in a position to aid her financially, but he refused to do so. Saying little in public about her disagreement with her husband, Matilda Weaver secretly took a job as a maid in a motel on the outskirts of Boston. She was able to save some money, although at fifty cents an hour it amounted to very little, and turned it over to her daughter. In doing this, she made Grace promise that the money would be spent only on the children, and that a doctor would be present at the delivery if Grace refused to have the child in a hospital. When a baby girl was born, Grace told her mother that she could never fully repay her, but as a start she named the child Matilda Cleveland Weaver. The older woman cried on hearing this, but her husband's anger never fully subsided.

Grace Weaver would have three more children. She would be married twice. The first was to a perfectly wonderful man who became incensed at the prospect and then the actuality of unemployment. For several years he had worked in a tire factory, but when the factory was shut down he lost his job. For months he hunted new employment, but the only available jobs were that of janitor and night watchman. Unable to bear either prospect, he began to drink. Gradually, too, he took to leaving his wife, his own two children, and the two children Grace had brought to the marriage, for increasing periods of time. Finally, on returning after an absence of three weeks, he left for good.

Two years later Grace married Randolph Potter, a steward on the Pennsylvania Railroad. Widowed, and the sole custodian of his three children, Randolph needed a wife. His first look at Grace Weaver Williams, for she carried her first husband's name even after he deserted her, told him that this was a loving woman and a decent mother. A justice of the peace married them. Everyone moved into Grace's apartment, and so, on their first night as husband and wife, the Potter family consisted of six people. Ten months later they had their first child together, a son, Raymond Maynard Potter.

Matilda Cleveland Potter, her last name having been changed, grew up in Boston. The name of her real father was never divulged. Everyone could see that she remained her grandmother's favorite. She was always a happy, pleasant child. The coming and going of people around her and the almost annual shifts in her family situation seemed to have taken little toll. By the age of twelve, it was apparent that she would be a handsome woman. Her body was filling out. Her skin was smooth, and her long black hair had the same silken texture of her mother's. She walked like her mother, too, a fact that annoyed her grandmother. "That girl asks for trouble with each step," the elder Matilda told Grace one evening while the children were sleeping. Middy, as she now was called by everyone except these two women, had been closely watched coming home that night. A group of young people had been playing somewhere, and Middy had returned sweaty and dusty, but extremely happy. She was wearing shorts and a thin summer blouse. Both women were perplexed by the child's appearance and mood.

School continued to trouble Middy. She barely got by, but despite the pressures she always seemed happy. On this spring evening, several weeks in advance of her thirteenth birthday, however, the two older women worried about her and the precocious sexuality that gleamed in her appearance and movement. Grace was herself about to be married to Randolph, an event that seemed to please Middy, although the youngster spoke little about it.

Middy's thirteenth birthday was celebrated by the members of her family and some school friends as well. Three boys, especially, caught the attention of Grace and Matilda. Randolph, who had brought his three children to the celebration, looked on with a fatherly pride. His own thoughts were on Grace, who that day looked especially beautiful. With his arm around his bride-to-be, he whispered, "That there's one gorgeous young lady. If I had seen her first I never would have been interested in you."

"Don't joke about that, Randolph," Grace came back. "I'm not so delighted about the way she looks. Look at those boys over there. They got their tongues hanging out so far we may have to call the police in here before this thing's over."

"Aw, go on. You just forget what you were doing and thinking at that age."

"That wasn't nice," she responded, pushing him away.

On the afternoon of her mother's marriage to Randolph Potter, the children of both adults convened at Middy's house. As the oldest, she and her brother Robert were to watch over them. Outside the rain poured down, the sky looking as though it would never clear.

In the middle of the afternoon Middy received a call. Her brother answered the telephone. "It's Farmer. He wants to speak to you. Jesus, you two oughta be getting married what with all the time you spend together, and the phone calls and everything." Middy dashed to place her hand over the phone and shouted at her brother:

"Shut your mouth! Go away and let me be." Her tone with Donald Farmer was considerably different. Although they did not officially date, the two had been seeing each other quite often. Middy thought she loved him. When they began to kiss and touch each other she told him so. Donald Farmer never said anything to her about love. Over the phone he asked to meet her.

"How about Crusty's house?" he suggested. "His mother's not home 'til six." Middy refused him at first, describing the day's events and how she and her older brother were supposed to babysit for her half brothers- and sisters-to-be. But at sixteen, Don Farmer was persuasive. Middy convinced Robert she would return momentarily. She ran out into the cold rain and met Don at Crusty's mother's house, the rain having saturated her clothes. That afternoon they made love. It was the first time for Middy. On returning home, she concealed the excitement she felt. There was a part of her that ached, but even the pain was pleasurable. Six years later she would recall that day and what she would describe as that "beautiful love ache." She would recall too the months that followed as hurtful months, but as "the days that made me what I am today. If I could look at a time and say, I am today because of this thing or that thing, it would be those months."

My own friendship with Middy began when we were introduced by her special boyfriend Thomas Jackson, a young man enrolled in a local university. Knowing of my research interests in life histories and people's involvement with time, Tom insisted that I meet her. Middy agreed to speak with me with an understanding that I wanted to learn of her history and present circumstances. There was nothing more, although in the beginning she joked about my being a psychotherapist. I assured her that therapy was not my intention. After all,

I argued, I had come to her. In time our discussions focused on her family, her childhood, and her prospects. Always I marveled at her poetic ways; always I felt helpless at my inability to call on experiences of my own youth that might have helped me to feel what she had experienced.

"See, the thing was that when you're that young, you think every little crush or infatuation has to be love. That's the way it was with that boy Donald. Anything he told me to do I did. And let me tell you that he had what I thought were some pretty weird ideas. But I liked it. I liked pleasing him. I had never pleased anyone like I could him. So I went along. That first time, like I told you, I didn't know what I was doing, but I got better. I remember being surprised when he invited another boy over to his house and said that this other boy liked what he liked. I was just doing it for Don. Well, then, I don't know. You don't want to hear all the details anyway. The thing was that gradually I met up with girls who I talked to, and they said I should be getting paid for doing it; and then I met two guys who said they would help me get started, which they did."

"How?" I asked. It was late afternoon. Middy and I were eating hamburgers together in her Boston apartment. One entered the apartment through a small hall, and then walked through a small kitchen to a living room. Beyond the living room was a bedroom barely able to hold a double bed and dresser. The apartment was neat and gaily decorated. Middy had always taken great pride in it. In the bathroom, all sorts of laundry were hanging up to dry.

"Well, for one thing," she answered my question, "they pay all my rent and buy me lots of clothes."

"Is that right?"

"Yes, they do."

"And what, may I ask, do they get in exchange?"

"Not what you're thinking." She smiled at me. It *was* what I had been thinking. "You want another coke?"

"No, I don't think so."

"Well, I do." She rose from the couch and walked to the kitchen. "They get 50 per cent of everything I make and all of the money

I get for the room." In the area of Boston where Middy works, the price of prostitution is usually based on the rental of the room as well as the sexual act. The couple usually negotiates price, with the woman constantly trying to manipulate both figures in order to make it seem as if her client were receiving an especially fine deal. The room, in Middy's case, is her own apartment. The men who take a percentage from her own the building. Of the few prostitutes who live there, some are forced to pay rent. For the present, Middy is subsidized because the money that she brings in exceeds what a monthly rental would come to.

Middy had returned with her drink. She plopped down on the couch. "Well, what do you want to know about me today? I've told you everything about my family, such as it is. You've met my mother and grandmother, and Robert. So now you have to ask."

"Well, that's kind of tough, Middy," I began. "I guess I want to know more about your life. What you think about, what matters to you. What your work is like." It is difficult to ask these questions and not sound like a psychotherapist.

"You want to know what's in my mind?"

"Well, yes, to some extent, I suppose."

"I will tell you everything I can. Just tell me when it's seven, though, so I can start to get ready. When *you* go home for dinner is when *I* start working, you know. Tom probably told you that." Tom Jackson was at present her favorite boyfriend. He is a powerful and articulate young man, headed, it now seems, for a career in law, perhaps in politics as well. His energy, intelligence, and good looks have captured Middy, but she fears in time that she will lose him. Middy is as lovely as Tom had described. Tall, attractive, beautifully dressed, she appears to be a fashion magazine model. The image of our first meeting seemed utterly incongruous: Middy elegantly dressed, munching a pizza. Every time we met thereafter she wore different clothes, often changing three or four times a day. Clothes meant a great deal to her, and so it was not surprising that she would frequently speak to me about them.

"Okay. I'll keep an eye on the time," I said.

"That's a funny expression," Middy responded. "'Keeping an eye on time.' I play funny games with time, you know. Like I'll pretend that I am young again, or then very old. I don't like the thought of being old. I always think about it when I've seen my grandmother. She reminds me of what it's like to be old. I kind of shudder inside. It makes me so nervous the thought that I will die sometime. I sometimes think about that in picking out clothes. I think about how old I would like to be and then buy something that reminds me of that age. Like this blouse." She fingered the sleeve of her shocking pink blouse. "When I wear this I'm ten years old. It seems silly, I know, but when I see this blouse I think to myself that if I wear it I will be ten. When I was ten my mother had a boy friend who bought me a dress. I remember he took me to Filene's and asked me what my favorite color was. Blue was my favorite color then, but I didn't want to tell him. So I said pink and he bought me this dress. I think I wore it only once or twice. I couldn't stand that man. He kept hanging around our house. I didn't like anything he gave us. I have a yellow pants suit. You've seen it?"

"Yes, I have." My face told her how marvelous I thought she looked in it.

She grinned. "Well, that's my thirty-year-old outfit. I'm someone very mature and suave when I put that on. Sophisticated, too. I even have a different way of walking with that because I imagine that I'm a very wealthy woman with all sorts of servants, and that I am preparing to have a big party in a garden. All kinds of famous people are going to come. I can see the faces of some of them. Movie stars and people from television that I can recognize. Then there are my jeans." Without looking back, she pointed over her head to where a pair of blue jeans hung over the back of a chair. "Those are definitely my childhood." She smiled. "But you want to know something very interesting? I can almost imagine that I'm not only younger but that I'm also somewhere else in my jeans."

"Like where?"

"Well, it's hard to say. I can't really tell. It's almost like I'm in a tiny village with lots of fields where you take long walks with people and never see anyone else. I imagine walking with a tall man. I run and jump and play with the leaves. And there's a church. There's a tall white church with a bell in it. The church has a clock and a bell but the clock never moves and the bell never rings. It's really weird." She paused a moment. "I never dream. Once every so often I have these crazy dreams, like, but usually I never dream when I'm asleep. Maybe it's because I have so many of these other thoughts that I don't need to dream. You think maybe that could be it?"

"Perhaps. I don't know." I think at times we both might have wanted me to act more like a therapist. I know I felt certain kinds of utterances coming up inside of me. Moreover, I believe she wondered about my "professional" reaction.

"Me either. Well anyway, I see this man and we're walking, and I guess that's all. He seems very young though, although I imagine that maybe he's like my father. But he always seems to look too young. My grandmother always says that about me and my mother. 'We look too young,' she says. 'It's going to get us in trouble.' I don't know. Maybe she's right. I don't feel young, though. That's always something I wonder about. When I meet my friends I seem to be one age. With my family, the little that I see of them, I'm definitely younger. Like when I'm with my mother she always makes me feel as if I were five or six. Or even younger maybe. I think she would like me to be young, maybe even be a baby. She always used to tell me how much I looked like her, or how much she looked like me. It made me feel funny. It was like if she kept me young then she wouldn't be getting any older herself. It bothers her that she's getting older. She never really talks about it but there are little ways that I can tell. Like, when she'll put on a dress or slacks or something, she'll look in the mirror and say, 'Not too bad for a lady that's getting on.' Then if I'm there, I'll say, 'Mom, you're still pretty young.' And she'll say: 'I'm no spring chicken. God's just teasing me by keeping me looking young. Maybe he'll give me one more chance to make amends.' She's had it tough, my mom. We fight a lot. She's very angry about my work. My stepfather won't even talk to me about it. Mom is begging me to quit. She keeps telling me what it's going to be like when I get older and have nothing. 'Wait 'til you're forty and you're beat up,' she always says. I just can't imagine myself that age, or fifty either. I can see myself very old. There's a little girl in my mind I think is probably me when I was small. But that in between time is like it's all empty. There's nobody or nothing there to fill it up." Middy looked at me and grinned. "Don't go taking any offense now."

I smiled back at her. Her images and recollections were so rich, there would be enormous problems remembering all that she said.

"You remind me of one of my customers. He's about your age. It's very strange being with him, too. I always hope he might come driving around. He'll put on his jive for me, and tell me how pretty I am. Then we'll come up here and I'll start taking off my clothes and he'll just sit there, right there where you're sitting, and start talking about what he's going to be someday. He's got all kinds of plans. Ooeee, there's not twenty people got the imagination he's got. He doesn't ever promise me anything. I'm not in his future. He just sits. One night he came up and was sitting there in his shorts. He still had his shirt and tie on. And one shoe. I was standing over there near the kitchen naked like a baby, and he was spinning out this future, like how he's going to be president of this great big company and have limousines pick him up in the morning and take him home at night. He didn't even see me standing there. Must have been twenty minutes went by. And all the time that he talked of his future, I was trying to put together the pieces of my own life. It was one of the only nights I can remember when I got paid just to listen to a guy talk. Strange dude, but I like him a lot. Maybe more than anyone else I've met. I refused to take money from him, but he insisted. So I took my money, but I remember I didn't go out again that night. I just lay on my bed dreaming about what I usually always dream about."

"Which is what?" I asked.

"Oh, I don't know. I try to imagine what my father looked like. Was he big or small. How much he weighed. How come it was I never met him. Things like that. You can imagine them as well as me. But there *is* something special about him—this man I don't even know. I always think that I was destined to become a prostitute. I won't do it that much longer, of course, but for now I always think I had to, that maybe it would be the only way I could find my father. Like one night I would be making it with some guy and then I'd know. All of a sudden in some way I would just know that the guy was my real father. Pretty queer, huh?"

"I don't know. Not really, I suppose." As a therapist, I would have jumped at her question; as her friend, I preferred to remain silent and reflect on her history.

"I hope I'm not sick to think these things. But I do think that if I could just meet him, my life—I don't know how to say this—that my life would fit together better. It's like, well, you wouldn't know, which is good, 'cause it isn't the best of situations; but when you don't have one of your parents with you, or if you don't know one of your parents at all, like your father, then you don't really have a whole life. It's not only the past that seems screwed up, it's the future, too. Like, when I'm dreaming, I think that my future can't get going until my father comes, or until something way back then has been settled. It's always a feeling of being, like, going nowhere, although you're making a lot of effort to move. And you can't move, no matter what you try to do. Sometimes I feel like time was moving sideways and I wasn't going to have to get any older for a long, long time because the time that was passing was not making me any older; it was only filling up all those empty places in my childhood. I'm getting confused. I'm starting to sound like Tom when he's studying for his exams. He gets so pent up he doesn't know what he's talking about."

"You're doing fine, Middy," I responded. "If you only knew how extraordinarily well you are doing." I could only think of the occasions when I had heard people speak of the intellectual poverty of persons like Middy. If only they could hear her.

"Thank you. I'm doing the best I can. It's just that so much of my feelings I can't always find the right words for. Tom would be good at this. Have you ever asked him to tell you about his life?"

"I have, yes. It's a very different life than yours, isn't it? I can't always believe that you both are as young as you are." Even as I spoke my remark sounded patronizing to me, but Middy heard the flattery and admiration I so much wanted to communicate to her. She shook her head from side to side.

"I'm not really," she said. "In so many ways I'm very, very old. I think about death a lot. I think that someday I will be in a place that will be totally empty. There won't be anyone there. I'll just be calling for people and no one will be there. And I'll feel time rushing by me. Like the wind. Time will be the wind blowing on my face, faster and faster, harder and harder, 'til I won't be anyone anymore. I'll be in the ground with all the dirt and stones on my face and everything. In a box. It will be dark."

Having been caught up in the language of this nineteen-year-old woman, I had not seen her begin to cry. Suddenly, there were tears falling from her eyes in a peculiar way. They seemed to line up on her lower lid and then, as heavy droplets, they fell on her chest one by one. Yet strangely, her eyes were never blurry or teary. They stayed as clear and pure as they had been all afternoon. She must have felt this, for she never reached up to wipe her eyes. She merely looked down at her breasts, and with the back of her hand spread the small water marks that fell on her blouse. And she never once stopped speaking.

"You know, what I've been thinking about is that if a person has a good time when he's young, then maybe he isn't afraid that he's got to die. It's more than living a full life, which is something my grandmother always crams down my throat. Of course, now she's telling my stepsisters. It's more though, 'cause you can't have a full life when the beginning gets all messed up. I can't look forward like Tom does. That boy has more plans than you could shake a stick at. He's got every little step planned out. He always says that if you took his routine away from him, though, he'd fall over like a dead dog. Me, I got a whole mess of routines. If you stay around long enough you'll see me go through one."

"Which is that?" I wondered aloud.

"Making up. I have to do it the same way each day. First my hair, then my eyes, then my mouth. Then, just before I leave, I spray my hair. It's like things won't work out if I don't do these things and do them in the right order. Tom has the same thing."

"Me too. I always shave the same way: first the left side, then the right, and then under my chin." Finally, I had something to share with her.

"Hey, maybe the three of us are related somehow." She was smiling.

"Well, maybe. You never know." She could see that I might have been hurt by this feeling of Tom and her being separated from me.

"Well, I think we could tell pretty easily, in that you're white and we're black. Want to know something?"

"What's that?" I hoped the subject might change, for like them, I have nothing to say when I am told that race will keep me apart from people I care for.

"When we first met, I mean, before we met and Tom told me about you, I tried to imagine what you looked like. I never met anyone who was doing work like you. Just talking to people. Then I thought, maybe he's found my father. 'Member I laughed when we met?"

"I sure do. I thought maybe I'd forgotten to finish dressing or something."

"Yeah. I remember you checking yourself out." She laughed aloud for the first time that afternoon. "I was only laughing then 'cause I thought maybe Tom had found my father, and when I saw you the whole idea struck me funny. Ain't no one like you who made me what I am. I think I can be pretty sure about *that*." Grinning, she looked at me. "Real sure." She paused. "Where was I?"

The exchange was typical of our conversations. Middy recognized my insecurity with her and the times I felt distress over some issue or idea. She sensed as well our respective flirtations with material that we might have believed to be appropriate only for psychotherapy. She knew, in other words, that my distress derived from racial or social-class differences and from our temptations to deal interpretively with what she spoke about. But she saw me controlling this temptation, and as recently as our last visit she wondered whether she should visit a psychiatrist.

"With routines and patterns," I reminded her.

"Right." She said the word softly. "I have lots of routines, like you call them, things I have to do all the time to make things work out, and stuff. When I was a child, I remember thinking that maybe there was something I would have to do to make the night come. Then, when I'd get into bed, I'd think, what do I have to do now to make the night go away again? It was real weird. I liked the idea that maybe I had something to do with the night and the day and each changing into one another. Now I think about those kinds of things a lot 'cause I don't really see that much of the day since I'm always sleeping. It's strange being awake when you know lots of others are asleep. Sometimes, when I go out at night, I think of my mother and what she's doing, or my grandmother and what she's doing. Sometimes, but not always, I see myself doing the same thing. I'll be them. I'll be old and wondering how much longer I have to live. It's a strange feeling, almost like a dream, you know?"

I wanted her to speak forever and hoped I would manage to remember even bits of what she said.

"Yeah, I'll be there and here, I mean walking out of my house and turning out the lights, and thinking if they're all right, healthy and everything. Then I'll have this feeling that maybe I should go into my bedroom and see whether there's anybody in there asleep in my bed."

"Who would be asleep in there?" I asked, as any researcher would. As any therapist would, too.

"I don't know." She began to laugh quietly. "Well, sometimes you have to make sure no one is left over from the night before." I laughed with her. "But maybe I think I'll find myself in there; an old lady all shriveled up and ugly, and I'll be dead. Sometimes I can't even get that idea out of my head. So I laugh it off and leave. I kind of start right over there at the door," she pointed, "near the light switch, you know?"

"Yes, right."

"And it will suddenly be like I *am* old, at the end of my life, like right now was lifted out of place and put way at the end. It's like time just rushed by real quickly with no interruptions, you know, like a train going very fast and not stopping anywhere even if people want to get on. It keeps going until it stops, and suddenly there you are somewhere." We waited in silence together. Middy seemed unconscious of my presence. Then she added: "And off, too, I guess."

"What's off? I didn't get that, Middy. What's off?"

"I just thought that if time is like a train and it keeps rolling along, then it not only doesn't let people on, it doesn't let people off either. Boy, this is getting weird."

"I don't think so."

The irony and presumption of conversations like these is that one wants, truthfully, to hear of the sad or poignant moments in another's life along with the joyful and pleasant ones, but prays that the telling of them will not make the person especially sad or lonely. Perhaps it is unfair and injudicious to arouse sadness or fright or loneliness by one's questions; then again, it is a human gesture to allow another person to speak, to reveal the outlines of her privacy and guarantee no interpretation. What tips the scales between indelicate intervention and gentle acceptance is how, finally, the teller is honored by the work and ethics of the one who listens.

"Well, then, okay." Middy looked at me and smiled. How, I wondered, did her clients feel when she smiled at them. As good as I now felt? "Now let's see, what else can I tell you about all this?" She rubbed her index finger across her lower lip. "You know what it is, this time thing? I never feel that the present moment is where it's at for me. I never say to myself, 'Now this is the moment right now and this is where you are living.' It's like my body knows that now is now, but my mind is always running all over God's country. Like, I can make myself be three or fifty just by saying the numbers to myself."

"That's like you were saying before."

"Am I repeating myself too much?"

"No. No. Not at all."

"I do that a lot, I think. I'll say something and then a little later I'll say it again. And someone, usually Tom, will say, 'You just said that,' and I'll be thinking now I *know* there's something wrong with my brain. This time I just know it."

"Well, repeating things is like having patterns or routines. You know, like my shaving thing."

"And like my dressing thing. But when I repeat things day after day, I think what I'm doing is making believe that time isn't passing at all. Either that or I want to make sure that it *is* moving. Oh, that's another thing." She looked intently at me as we both adjusted our positions. "This you will find interesting, I hope. Sometimes I think I can stop time; you know, I can live a little while without anything really happening. When I get up in the afternoons, I'll say things, like, 'Well, another day,' or 'Here's the new day, better go catch it.' But those are just expressions, they don't mean anything. I'm not *really* feeling that the days are passing and that I'm getting any older or anything. Even my birthdays don't make me feel that I'm *really* getting on yet. Like my mother, when she has a birthday she goes crazy worrying that pretty soon she'll be ready for the scrap heap. But I always feel young. Young or real old, like.

"But there's this feeling that I can make time stop. It always happens when I'm working. I'll be with this guy and we'll go through all the jive on the street, you know. He'll say he's coming or he's not coming and we'll mess around. Then he comes up here—if I'm using my house. Sometimes I don't. Then we'll say a few things. Usually they ask me whether I got booze or drugs and I'll have to tell them real seriously that I run a good business and don't mess around." She spoke to an unseen client: "'If you don't like what I'm selling, go somewhere else. There's a nice bar over at the corner of Columbus and Mass. Ave.' So then I'll start to take my clothes off so he can see I'm not about to be wasting time, and then it's like all the clocks in the world stop. No matter what time it is wherever you would be, they stop." Middy no longer looked at me.

"I see myself almost every night like that, standing naked and real beautiful, and nothing anywhere is moving, and no one is going anywhere. All over the world people have stopped breathing. They haven't died or anything like that; they've just been stopped for the moment while time stops. And I stand there very still and let the guy look at me. They love to look at you. I think some like it even more than touching you, or, you know. They look at you as though they've never seen anybody naked before. Most of 'em are just plain tired out, I suppose, from looking at their wives. I can usually set the lights up so that my body kind of shines in the dark. I don't even hear the traffic then. Everything is silent and soft. I'm not thinking very much about the guy. It's like he's not even there. I try not to think too much about him. Then I'll walk past him, through this room, after I wash, and make believe that I'm telling the world, 'Not yet. It's not yet time for everyone to get moving again.'" She held her finger up to her mouth. "Shh. Then we do what we do, and when I get dressed and put the lights on again, I'll say, 'Okay. Let's start the clocks,' and everything comes to life again.

"I think that's the best part of what I do. Folks talk about ecstatic moments. I know the guy feels all good. I make my men feel good. But sex for me, except when I'm with Tom, is just a chance for my mind to get free. I can have one of these guys be whispering all sorts of things in my ear, and inside my head it will be like a movie. I'll see myself at different ages. I'll be in school, where I did horrible things, or I'll be in the bathtub with my brother, and my mother will be washing us. Far out, huh?" She did not wait for a response. "Or I'll be with my grandmother. Like my grandmother will be holding me on her lap and rubbing my back or reading a story to me. She read to all of us when we were kids. But it's funny though, as much as Tom and you speak about them, I don't like books. I can't get by the first three or four pages. If you write a book I'll read it though. And if Tom ever writes one you can be sure I'd try to read it."

"I know you would."

"I would."

"I know."

"Anyway, we're all naked in the bathtub and it's warm and nice, or I'm on my grandmother's lap and that's warm and nice, too. What you should know is that when I think about these thoughts, when I'm remembering, I really feel as though I were right there. It's not like I can be in this time and that time at once. I'm really there. Believe me on this. Time just picks up where I guess it left off once. That reminds me of something else. Aren't I a good person to interview?"

"You really are."

"I should charge you."

"I've got no response for that." We smiled at each other. "I like you a whole lot, Middy," was all I could manage to say.

"Thank you. I like you, too."

"Good."

"What was I saying? Oh, that time picks up. What I make myself do is imagine forward or backward. Like I was saying before, I can make the future, although I don't know exactly where in the future it is that I am, play before me like on a movie screen. Whole scenes go past in front of my eyes, even if I keep them open. The same is true for the past. Long scenes, like long memories, go on and on and on. They come back, or it's like I go back there to them, like they've never ever gone away. Do you know what I mean?"

"I think so." Was this really the first time she had collected her thoughts about time?

"It's hard to explain. Tom could explain it better because he has a better vocabulary. It seems when I hear myself talk—and repeat things like I do—that I keep using the same words over and over again. That's true, isn't it?"

"I don't think so. I haven't been thinking about it."

"Well, you'll just have to suffer through it, no matter how bad it sounds. I know you know that I never finished school."

"I know that."

"Well, I thought that just in case, you should know because it shows why I don't sound that intelligent, like you and Tom."

"You make me feel very uncomfortable, Middy, by saying that. Truthfully, I've been sitting here thinking what an extraordinary gift you have for describing your life."

"Well, maybe I do. I don't know." There was an unusual modesty about this young woman, as well as a certain diffidence. It seemed that if she were to take a single step out of line, something serious might result, something for which she ultimately would feel responsible.

"Okay, I'm sorry. Let me tell you the rest of that other thing. See, I can get these memories, or whatever you would call them, and have them strung out in this long line." I could see her searching for ways to express her thoughts. "Long, long memories or the future things take over and stay around inside of me. Then I'm really a part of them. I can't even get out of them when I want to. It's not like the way I stop time, and then start it again. This is different. I imagine I can see whole weeks or months of when I was a kid, but most of the time these weeks don't fit together. They don't even have beginnings or an end, you know. They just keep running on . . . disconnected. They're disconnected! It's like there are spaces between them . . . these . . . big spaces. Like I'm missing parts of my life, or like there are holes all over the place and I don't know whether those are supposed to be the times I don't remember, or the times I just don't want to remember for some reason. Do you think that's it?"

"I don't know, it just might be," I said. Again I was uncertain how to respond.

"There are some times, I know, that I don't want to remember."

"Like when?"

"Like when I was young and my mother's husband left us. I

remember that day real well. I try all the time not to remember it, but that's silly 'cause there's nothing I can do to forget it. I pretend that I can change it but you can't get rid of thoughts like that, can you?"

"I guess not."

"I remember all the fights. When my mother would get sick, I would wait on her and get things she wanted and talk with her. Then she'd say something or do something and I'd get angry."

"Like what kinds of things would she say?"

"Oh, I can't remember exactly. She would tell me about a whole lot of things I didn't really want to know about. Like her marriage plans, or guys running away from her. What was happening was that instead of *me* being in bed being waited on by her, *she* would stay in bed all day, even if she weren't that sick, and I would have to wait on her. Not only that, it would be like she was the little girl and I was the grown lady when all the time I wanted to be the little girl and have her be the grown lady. My grandmother was never like that. Not my grandfather either. He doesn't speak to me at all now since I've been doing this." She made a gesture as if to enclose the space of the apartment. "But my mother was always the odd one in the family. When she'd talk to me that way, about all her personal problems, I always felt I was being bad for not wanting to listen. Guilty. I felt guilty. I'd try to talk myself out of those feelings but it never did any good. It was like she wouldn't let me be a little girl even when you're supposed to be a little girl. She always wanted me to be like her sister, I guess. She wanted to confide in me and tell me everything that was running through her head. It made me mad because I didn't want to grow up as fast as she was making me grow up with her telling me her problems and stuff. Like you would say, she was taking my childhood away from me."

"I would say *that?*"

"Yes. Don't you remember one night at Gradie's with you and Tom, you were telling him how many times you saw people biting away at someone's childhood? It was something like that." I remembered what Tom Jackson and I had been speaking about. I noticed, too, Middy's way of combining all the moments we had spent together to form that discontinuous but still prolonged movie of time about which she spoke.

"I did. You're absolutely right. I remember now," I answered happily.

"See. It's like I said, I can get the pieces together. You know what it is? People bring me together with different parts of my life. Like the past and the future are either there or not there depending on whether I imagine myself being with people. Like, time is definitely moving when I'm with my mother. When she was sick it really dragged though. Nothing I did could make the days go faster. But when I think of scenes that have no people in them, then it's like time is passing me by."

"But time stands still, you said, when you bring a guy up here."

"Oh, that's different." Middy stared at me. Then quickly her head moved as she took a visual inventory of the room. I imagined she wanted to make certain that nothing was missing, that I had not accidentally taken anything from the room. "That's different 'cause I'm not really conscious of the guy. When we're still in the street, sneaking around, you know, I sort of hypnotize myself so that I'll be able to forget that anyone is going to be with me. It would be nicest of all, sometimes, to be . . . what I am, to just be by myself. But of course I know that's silly. So I just pretend that he's not there and then time is dead, or at least I pretend it's dead."

"Middy, this may seem like a strange question. I'm not even sure why I'm asking it, but, can you imagine a clock with a face on it? I mean a human face on it . . . ?" She had already interrupted me.

"You mean like a man's face?"

"A man's face, a woman's face, a child's face. Is it a strange question, I wonder . . . ?"

Again she interrupted. "A man's face. It's my father's face. There's no doubt about it. I know in some way I can't tell you about now that I know what his face looked like. It's his face I see. I could make myself see him now if I really wanted to. That's right. That's what I was trying to tell you a minute ago. I think that's the reason that my way of thinking about the past and the future, and I suppose the present too, all relate to my mother and father, the reason that all my thoughts and pretendings don't come together. I mean connect," she looked at me and winked, "is because he's missing. He's the important missing piece in my time business. That's why, too, I can't be good friends with time. Time's my enemy. Like I say, we can't get together, time and me. I know that I don't live in the same way that anybody else does. Even the other girls. It's not just the job I'm thinking about. It's my background that makes me special. You can't get rid of your background, you know, just 'cause maybe you'd like to. Old man time won't allow me to. He won't allow anyone to. I'm pretty sure of this 'cause I've tried. I used to do drugs, thinking I could get away from things, or maybe forget a little here or there, but you know how that stuff always turns out. Even when I'm in bed and have some big ugly guy killing me to death, I don't even see him. I keep my eyes closed, and you wouldn't believe the kinds of things that cross my mind."

"Like what?" I asked. It was, truthfully, a question I had wanted to ask her for as long as I could remember. What *do* "they" think about?

"Everybody asks me that. Even the police. We go, Tom and me, drinking with some guys on the force that we know. One of them takes a little on the side from one of the sisters, in case you wanted to know. Anyway, he's always asking me, 'Baby, what's your mind doing when you go balling some dude?' I tell him I'm just thinking about what a good time I'm having, but that isn't the truth. Actually, it's usually the same sort of things. Say, what time is it? Is it seven yet?"

I had forgotten about my promise to remind her of the hour. "It's almost seven," I said, after walking into the kitchen to look at the clock on the stove. The time had passed quickly. I had been sitting with her now for almost three hours.

"Okay," she said, "we've got a few more minutes. But you're going to come back anyway, aren't you?"

"Of course."

"I'm relieved, then. Well, what do I think about? Sometimes I think about that crazy Farmer boy and the first time I ever had sex from a guy. We used to do it on the roof of our house and I used to always think we'd roll off and someone would find us there on the ground. Then I'll have a thought imagining how my mother looked when she was a child. There are some pictures around the house but they don't really tell too much. Then I'll be her, making love to my father, or some guy I know is my father. Those are the times you'd be interested in because when I imagine that stuff—I haven't ever told this to Tom, I don't think—I see everything as though it were really early in history."

"You mean hundreds or thousands of years ago?"

"No, no, not like that. Like I'll see my mother dressed like they used to dress. I guess it must be from television I get all these nutty ideas. Or I'll see myself in a basket, as a baby. I can barely peer over the edge of the basket, and there are all these rattles and toys, and I see her making love to some guy, and I know, even though I'm just a baby, that I'm not supposed to be seeing what I'm seeing."

"Are these dreams you have at night?" Suddenly I sounded as professional as I ever do.

"No. Like I said. This is what I see when I got a guy here. It sounds pretty weird doesn't it?"

"No, why?"

"Well, you're asking me whether it was a dream, and it isn't."

"No, it was just that I was wondering whether people could remember such things, and I'm sure they can."

"Well, I don't know whether I am remembering these things or making them up. I'm not asleep, you know. If I were," she began to smile, "I'd never get my money."

"I guess that's true. Well, what else?" I wasn't certain that I wanted to hear more about this. I was beginning to feel uncomfortable.

"Let's see. I think about food a lot, too. Bitter-tasting food, or ice cream and candy. Sweet food, mushy things, you know, like they grind up for babies. Or every once in a while I'll think that I'm getting sick to my stomach. Or, this is another. Sometimes I'll be lying there," she gestured with her head in the direction of the bedroom, "letting a guy do whatever he wants, and I'll suddenly be thinking of a house that I'd like to live in. I'll take, like, a room at a time, like a living room, and I'll think of what kind of furniture I want in it, and this color or that color, and what kind of curtains should be on the windows, and the carpets. Then I'll switch to another room, like the room that would be my bedroom, and I'll start all over again. Sometimes I'll really get excited by the way the room looks, and sometimes it will be ugly so I won't like it at all. Then I'll get angry or maybe feel disappointed. It's like I've made all these wonderful plans and figured out the way things should be, but when they get done, they don't look anywhere near as good as I thought they would. I even feel like crying when that happens. It's really disappointing. It's funny; it's like it really was going on." She paused, the look on her face revealing that her mind had not stopped working on that one idea. "Now this is something I once figured out. This is now me as a psychologist. I think I made this up myself but I might have read it in a book somewhere. In the first few pages, right?"

"Right." I remembered the reference.

"I told you that when I'm with people in my thoughts, that I'm on pretty good terms with old man time. But now here is something else. Naturally, the guy isn't invisible. I mean, I know he's there. Sometimes, it's even kind of nice. I don't tell Tom, but sometimes it is. The older guys are nice to be with. But I like young ones, too. Anyway, don't let me get started with that."

"Yeah, well first I have to ask you, Middy, am I old or young?"

"You? I'd say you're in between, on the way to old." She laughed.

"Okay. Just checking."

"In between," she repeated, studying me up and down and grinning like a little girl who might have just played a trick on someone and knew she could not get caught. "In between, yeah."

"What's your calculation? I got you sidetracked."

"Yeah. My calculation. Now don't laugh, now."

"I won't."

"When the guy is good and I'm enjoying it, which isn't all the time, my thoughts get involved with things like my plans. My future. That's when I start thinking about what I want to do in a few years, if I get enough money saved up. That's when I think about my pretend house, too. But when it's bad, when I feel sad or hurt, and stuff, like when he's hurting me and I might be afraid to tell him to stop, then I sort of close out the hurt by making myself think of things. It seems like those things take place a long time ago. I'm always a baby or a little girl then. Anyway, that's my calculation: When it's good I can think about some of the things that might happen to me. But when it's bad, when I feel that if I didn't have that big ape lying all over me I would probably be crying my eyes out, then it changes over, and I can't get out from thinking that my whole life is going on, like, a long time ago." She hesitated for a moment as if perhaps she had uncovered a flaw in her calculations. "No, that's right. That's the way it goes. Bad things make my mind go one way; good things make it go another way.

"Problem is that nothing ever happens in the bedroom that lets me think about the present, like, you know, this week. I always find ways that make me forget this week. Some people I know can't remember anything that happened yesterday. You ask them what they had for lunch yesterday and they can't even tell you." Middy giggled. "Me, I can't remember anything from right around me. I mean, I remember things but it always feels that I don't get into the rhythm of each day. It's strange. Maybe someday I'll change and be different. Then I won't talk this way." She remained motionless on the couch for a moment, her thoughts a million miles away. Then, shaking them clean, she rose. "I've got to get going. It's all right 'cause you're coming back, okay?"

"Okay." I was pleased that she would want me to return.

"I wonder about myself." Middy Rutherford walked from the room. "Don't go yet," she yelled from behind the bedroom door. I could hear her changing clothes and opening and closing drawers in the little wooden dresser. Occasionally, she would mutter something: "Stockings, nothing matches . . . Never clean clothes . . . Oh, look at that. Damn that stinking laundry. . . . What's this? This clean?" Then she would call out to me again: "Sorry, just part of the getting ready routine." When she stepped out of the bedroom, finally, the transformation in her appearance was breathtaking. From a young teenager, she had made herself over into a strikingly handsome and sensuous woman. Studying herself in the small mirror above the kitchen sink, she began to apply the make-up she kept in the single metal kitchen cabinet. Her actions were meticulous, and as she worked I could see her age with each stroke of deep red lipstick, aqua eye shade, and rouge. I heard her say quietly, "Now I'm nineteen . . . now twenty . . . twenty-five . . . twenty-eight . . . now I'm there."

"Where's there?" I inquired.

"Huh?"

"You said, 'Now I'm there.' I wondered where 'there' was."

"I don't know." She turned to look at me. "There. Just there. Nowhere in particular. Just old enough so that I'll be safe tonight in case there's trouble."

"You worry much about trouble or safety?" In this whole conversation I never thought of her in trouble. I never actually thought of her going on the street.

"All the time." She was teasing a curl near her ear. "Ain't no man in the world I can really be sure about." Never once during the entire afternoon did her voice swell with emotion. Now, for the first time, I heard anger tinged with the slightest morsel of fear. "Not one man, at least not yet."

"Middy," I started, "I wanted to ask you one more question. Can I?"

"Sure. Of course."

"Well, I've been thinking, as my kind always does, did you ever stop to wonder about the fact that you made love for the first time on the day, well, actually the very same afternoon that your mother got married that one time?"

She remained in front of the mirror. "Well, yes and no. I mean, it's more complicated than that. I didn't think about it when I went that afternoon, like I told you. Remember?"

"Very well."

"Well, I didn't think then, but I did a couple of years ago when I was looking in my diary and I saw that, they have two pages, you know, for each day? Anyway, I wrote about my mother's wedding on one page, and on the back page I wrote, you want to know what I wrote?"

"Yes."

Holding a curl between her fingers, she glanced at me. "I wrote, 'Mr. Farmer and I got married today if you know what I mean. It was good fun and I will do it again.' God, what a nut I was . . . And here I am doing what I'm doing." She returned to fixing her hair.

"So then you didn't really know about that connection?" The connection, that is, that I had made.

"No. But what I haven't told you exactly is that when I was ten and a half I was raped in my mother's bedroom. So it wasn't the first time." She inspected her face, pulling back her lips for a final check of her teeth.

"Jesus." I was stunned, but my reaction seemed to gratify her.

"I guess I didn't tell you that. I have many more things about my old past you have to hear about." I saw her move her tongue around her mouth as if to wipe her teeth. "A man came in one day to fix a sink or something. I was alone watching television and baby-sitting for a little girl who lived across the street. Maybe she's my cousin. Anyway, he did all the usual things to me. I didn't even yell much. I remember I was thinking I didn't want to wake the little girl. Later on I cried. My mother was furious. She called the police and everything. Let's go now." Middy took a last look at herself in the small mirror, sprayed her hair, hesitated for a moment, posing, then moved away. She turned off the living room lights and unfastened the chain lock on the kitchen door, which also served as the front door. We headed down the stairs together. Instinctively I found myself buttoning my top shirt button and tightening my tie. She watched me. "You look okay. Both my Toms look okay. When we get downstairs, though, you shouldn't go the same way I do. They got all sorts of police go buzzing around here. It wouldn't be so good if we got caught together."

"Okay," I agreed unthinkingly. We reached the front hall where a broken glass door led out to concrete steps and then to the street. Middy dropped a note that she had carried in her purse in the brass mailbox. The name on the box read "M. Rutherford." She saw me looking at the name handwritten on a piece of tape.

"Oh, so let me finish what I was telling you. They called the cops, and my mother tells them what's happened and all. They never found the guy, of course, but about two years ago I saw my mother in a bar one night with that guy. I know it was him."

"Are you sure?"

"I know I'm sure. It was the same guy."

"Did you ever say anything?"

"No. I didn't even mention it to Tom."

"Does Tom know about this, this thing?"

"My rape affair?"

"Yes."

"Sure, of course he does. You don't think I'd tell you things without telling him first. I have to tell him everything," she concluded.

"Love is nice that way, isn't it?" I smiled at her. She was altogether serious.

"Oh, it's not only love. I'm counting on him to change my life. He tells me that when he's finished with school he'll change the direction of my life. Then maybe I'll start to have all those things I can only think about now. Like I told you, I need other people to make me get together with the good parts of time, and he's the one right now. Him and Rutherford Hayes."

"Rutherford Hayes?"

"Sure." Middy had opened the glass door. "Rutherford Hayes. He's the guy I took for a name."

"This is all beyond me," I confessed, walking outside.

"It's simple. Tom and I like Elvin Hayes, the basketball player. So one night Tom was telling me about Rutherford Hayes and we decided that since we like Elvin Hayes I'd take Rutherford for my name."

"Yeah, I suppose that makes some sense."

"Now look, my friend." We stood together outside at the base of the stairs. The May weather was beautiful. "It doesn't make sense. The three of us know that. But this is the way you have to think. My whole life has been put together with pieces that never once made any sense to anyone. Not even to my grandmother. Like, why I was born, for a good example. So I figure that if I start putting things together now that *really* don't make sense, like, here," She opened the jacket of her pants suit slightly to reveal two powerfully contrasting colors, "then maybe it's possible that I can get my life untangled, straightened out, you know. Couple more crazy things and one good, really good chance, and I'll be out of here, and then, like we were saying upstairs, I can be on good terms with time, until I die. But that, none of us can do anything about, right?"

"Right."

"It would be awfully nice, though."

"To be away from here?" I asked. As we spoke, I felt the special tension that always precedes people pulling apart.

"Yes. To be away from here, and be gone; to be back there, maybe, again, and have it all straightened out. Like I was saying, just to have a chance to blow my breath on old man time. Who knows?" She stared past me down the street. Her look made me aware of the traffic and other people. "I'll kiss you good-by for a dime?" She grinned up at me as her reverie receded.

"Ten cents?" I said, feigning outrage.

"A nickel?"

"You're on." Clinging to the sleeve of my jacket, she reached up on tiptoe and kissed my cheek. The kiss was so much like Middy herself. "Someday I shall be a grown lady. You will see. I shall become

something. Someone, rather." I was about to speak, but she placed
her finger on my lips. "Please don't say you like me just like I am."
Holding her finger on my lips she kissed me gently again, looked
hard into my eyes, then turned gracefully around and walked toward
the corner where Newton Street meets Columbus Ave. I watched her
until she moved out of sight, expecting that at any moment she would
turn to wave or shout something back at me. But she never did.

Death, abandonment, loneliness, and a desire to be recognized
are part of what Middy Rutherford has built into her sense of time.
The freedom of her words and the expansiveness of her perceptions
permit her to become a child, an adult, an adolescent, and thereby
to freely move about on the horizons of time. Discontinuity in the
flow of her life seems to attract her, and special days, birthdays, death,
and marriage days provide her with a sense of evolution.

People are essential properties of time for her as well: "People
bring me together with different parts of my life." The mere presence
of others carries her forward or backward in time and allows her to
change, just as the loss of someone momentarily prevents her from
progressing. The loss of family members causes her to see herself
as standing still or drifting backward. Even personal appearance is
tied to time. For if one looks too young, one is too young.

In Middy's fantasies of the future, one detects the remnants of
her earlier identifications with her parents. To the unconscious, theo-
retically, a child is part of the parents, if not identical to them. And
so, we become our mother or grandmother, or ourself as a baby held
by our mother. From the perspective of time perception, the process
of identification intensifies our recollections of the past and stimulates
our fantasies about our parents' and grandparents' pasts. As others'
memories, then, become the content of our own fantasies and imaginings, we see ourselves living in their time and becoming the age that
seems appropriate when associating with certain people. These processes, too, can be observed in Middy's account.

The discontinuity in Middy's sense of time may reflect experiences
better for her left unsaid or unrecalled. But how extraordinary it is

that each of us can describe such discontinuities going hand in hand with our feelings of loss and abandonment. It seems natural that we might long to be a child, a baby even, if only to retrieve those hours when we felt wholly cared for and attached to someone, and existence seemed timeless. It is logical, therefore, that Middy dreams of her father while listening to a man design his own future. For in the sensuous touching of bodies, a man to a woman, a child to its parents, lies that feeling of attachment that each of us experiences as a timeless moment, a moment we continually seek to reproduce later in life. More than merely a metaphoric notion, primordial memory is what this young woman is able to describe: ". . . when I'm dreaming, I think all the time that my future can't get going until my father comes, or until something way back then has been settled."

Middy herself contends that she lives outside the present. Her words make it seem that she is able to fling herself through fantasy, memory, and expectation backward and forward in time, in the directions, again, of birth and death. This quality of apparently living outside time causes her to feel, ironically, that she is not fully alive. These feelings, too, are part not only of her self-conception, but of the role she assumes as daughter, woman, prostitute. No doubt, the shifting constellation of her family, and her own discovered guilt that derives from having witnessed that which she revives for us, plays into these perceptions of time. Consonant with the findings reported in Chapter 3, Middy indicates that pleasant experiences tend to drive her thoughts into the future, whereas unpleasant ones drive them backward, deep inside the past.

The research on young women's perceptions of time turns out to be particularly relevant for this exceptional woman. One senses a certain passivity in her perceptions of time in the form of a wish that others might carry her through life and eventually put a seal of reality on her actions and plans. Other people, moreover, must fill the empty spaces in order for the present to run smoothly into the future. Yet, none of this can disguise the hurt that she continues to experience and the knowledge that the future is unlikely to improve on the minimal gratifications she presently knows. It is the logic of time, the conditions of her childhood and adolescence, of poverty and migration, that converge to shape her wishes for her future along with her doubt of these very same wishes: "Couple more crazy things and one good, really good chance, and I'll be out of here, and then . . . I can be on good terms with time, until I die."

There is a natural flow of chronological time that underwrites whatever subjective perceptions of it that any of us hold. Our notions of order and routine testify to time's natural passage and to our desire to effect predictable ends and to repeat the gratifying actions of the past. Still, in our efforts to understand the subjective and objective properties of time, and how these properties are transformed during childhood and adolescence, we tend to overlook the effects of social structure, economics, and political realities on individual perceptions of time.

It is not merely symbolic to suggest that the empty spaces in Middy Rutherford's imagined temporal horizon, the spaces that she herself likens to death, are there because of circumstances dictated by society, as well as because of the unforeseen events that necessarily constitute any life. Through the roles allocated to us, and through our identifications and attachments to people who are especially important to us, we begin to form images of ourself and of our destiny. Their own discontinuous progression becomes our sense of incompleteness; their failures and experiences of oppression become our insecurity, caution, and unwillingness to trust; their youth is our old age, their death our missing or unsettled childhood. And just as people experience abandonment from being a member of a group that has never been afforded a legitimate place in the society, so, too, does a woman of nineteen conclude that time has passed her by, and that the future promises little hope.

PART
3

Socio-Cultural Perspectives

6
The Subjective Future
in the Context of Culture
and Social Structure

> It is by the meaning that it intuitively
> attaches to time that one culture is
> differentiated from another.
>
> Oswald Spengler (*The Decline
> of the West*, Volume 1)

As the conversations with Middy Rutherford reveal, images of the personal future are closely intertwined with daydreams and fantasies. In earlier chapters, we examined man's ability to distinguish realizable anticipations from unbridled flights of wish-fulfillment, and we have seen how images of the future gradually become imbued with a sufficient sense of reality to motivate present behavior, and to enable a person to resist the allure of immediate gratifications.

If only the present is real, then the realism with which images of the future are constructed will depend on the degree to which they are linked with and derived from the realities of the present—the process that we have called temporal integration. Few of Middy

Rutherford's dreams are relegated to the realm of "real" future events. She often feels that time is "going nowhere," in part because of her inability to connect the fantasies of her rich imagination with her current situation. They represent for her a different order of experience, not the real world of the present in its future extensions.

Images of the future will motivate behavior when they generate feelings of anticipatory pleasure or discomfort. To the extent that people avoid painful experiences as best they can, they are likely to be oriented toward future events primarily when their anticipations are pleasurable. When future prospects are bleak and felt to be subject to forces outside one's personal control, there will be little reason to defer any opportunities for immediate gratification and little basis for the development of long-range plans.

For all such aspects of future orientation, a person must ultimately rely on social interaction with others. Through his earliest contacts with society, a child acquires its particular symbols and constructs his personal images of time. In the social institutions that await his arrival and in the social roles through which he will come to define himself are the perspectives and expectations that inform his conceptions of the past and future and the nature of their integration with present experiences. The words through which a person expresses his hopes and fears for the future are stimuli to which others respond with approval or disapproval, understanding or incredulity. Their responses in turn will shape and transform the words and images through which he defines his own sense of self continuity.

The conceptions that Middy has formed of her past and distant future are profoundly shaped by her identifications with her mother and grandmother, and with the father she has never met. And the sex-role distinctions traditionally imposed on men and women in Western society deeply influence, as we have seen, the qualities of time that pervade their lives and the conceptions they form of their personal future.

To more fully comprehend the nature of time in human experience, then, we must explore the cultural myths and symbols through which time itself is construed. We must consider as well a person's social group, which rewards or punishes expressions of the future in different ways, reflecting its predominant value orientations and world views. In the complex social systems of highly modernized societies, we are also likely to find that social-class position and unequal access to educational and occupational opportunities will profoundly affect a person's sense of the future.

The normative changes in conceptions of the personal future that in Chapter 3 were seen to characterize development from childhood into adolescence refer at best to a limited segment of humanity at a particular point in human history. Implicit throughout that discussion was a conception of development that locates its underlying dynamics in the continuing interactions between the changing capacities of a growing child and the changing demands of his social environment. Psychological and social forces inextricably interpenetrate to shape a person's behavior.

If many young children tend to view their future as a realm of almost unlimited possibilities, this may in large part reflect their parents' tendency to raise them with a continuing emphasis on all that they may become in adulthood. Such a future-oriented socialization clearly presupposes an enveloping society in which time is viewed as above all pointed to the future. But in the broad sweep of human existence, that way of construing the dimension of time is relatively rare.

It presupposes as well that parents have found in their own life experiences that the future is basically trustworthy, that their plans may be realized if they expend the necessary effort, that deferred gratifications will come as expected. But such a view is likely to be maintained only in a society of stable social institutions and by people who have relatively ready access to opportunities for personal advancement. Children who grow up in the impoverished ghettos of urban America tend to receive rather different messages about their prospects for the future.

Part 3 explores the socio-cultural context of personal orientations toward future events. Our concern is with those external forces that shape and maintain a person's sense of temporal possibilities. The life study presented in Chapter 8 examines the expression of these forces in the personal lives of a working-class couple. This chapter explores the general impact of social class, as complex societies structure a person's access to opportunities and the conditions of his daily life in contrasting ways. Such conditions will underlie his sense of the future and the personal control he may exercise over it. We are also concerned here, but particularly in Chapter 7, with the impact of the social transformations that occur as modern institutions are introduced into societies that were built on a different set of basic assumptions regarding the nature of man and of time. In both of these pursuits, we shall be able to base our discussion on relatively solid empirical foundations in the form of systematic sampling and quantitative data. The excursion begins, however, on more slippery ground.

TIME PERSPECTIVES IN "TRADITIONAL" SOCIETIES

There exist today, although in rapidly decreasing numbers, organized social groups whose members live in relative isolation from the great currents of modern history. Their language remains unwritten, their livelihood based on subsistence agriculture, hunting, or fishing, in close harmony with the rigors and rhythms of nature. Do persons in cultures such as these tend to experience the flow of time and to conceive of their personal futures in ways that contrast markedly with the perspectives held by members of complex, highly modernized societies? Ethnographic reports suggest that they do.

Descriptions of behavior patterns in these relatively small, isolated, and homogeneous societies are based on anthropological observations together with intensive interviews involving a small number of informants typically chosen from among the oldest in the group. Their recollections of historical events, recountings of folklore and tribal myths, and recreations of their own life histories are carefully recorded. These comprise the raw materials from which the anthropologist attempts to build a "model" of the culture, to uncover the implicit assumptions in the world view of the society's members.

The impression is often given of a frozen uniformity across the population and of cultural forms unchanged for decades and centuries. Rarely are efforts made to sample systematically the population in question, to explore the diversity of individual attributes, and to speak in terms of frequencies and percentages rather than absolutes. This material should therefore be approached with caution, especially by one who comes upon it from a different discipline and with different purposes in mind. Ethnographic descriptions of temporal experience are nevertheless an invaluable source of insight into the mechanisms that appear to link a person's personal sense of time with central aspects of his socio-cultural environment.

A Non-Linear Time

Time is experienced only through the changes that occur within it, changes that are both repetitive and progressive. Human beings everywhere must surely know the periodic return of hunger pangs or the recurrent alternations of day and night, of activity and rest. They also witness the irreversible progressions of biological change, from birth through aging to death, or from seed to flowering plant. Is time, then, essentially a repetitive, cyclical, and alternating phenomenon; or is it, rather, progressive, linear, and unidirectional? The answer is not to be found in the realities of human experience, but in the cultural symbols and conventions through which all such experience is construed.

Marcus Simpson speaks bitterly of "the logic of time," the sense of its irreversible and rhythmic predictability: "Isn't a man alive going to be able to slow it down or speed it up." For Middy Rutherford, on the other hand, the days are not really passing: "Sometimes I'm convinced that I can stop time," she says. "That, you know, I can live a little while without anything really happening." There are cultural as well as personal forces that underlie such individual differences in temporal experience.

Edmund Leach (1961, p. 133) observed that time is likely to be represented by one of three central images. It may be viewed as a line, as a wheel, or as a pendulum. "You can think of time as going on and on," he wrote, "or you can think of it as going round and round. All I am saying is that in fact quite a lot of people think of it as going back and forth." Back and forth in a discontinuous sequence of repeated oscillations between polar opposites, such as day-night, day-night or wet-dry, wet-dry.

Seemingly irreversible progressions may be conceived to be parts of a larger, less visible oscillation. Civilizations grow for a while and then wane. Good times are followed by bad. Death gives way to birth, as the "soul" travels back and forth between the worlds of the living and the dead, or as rebirth follows symbolic death in the initiation of adolescents into the full responsibilities of adulthood (see Lévi-Strauss, 1966; Burton and Whiting, 1961). There are good reasons to expect that linear conceptions of time are rarely held by members of the "traditional" societies described above.

In the absence of systematic sampling and structured interviews, we can only infer from ethnographic reports the way most members of such societies experience the flow of time in their personal lives. Descriptions of "traditional" cultures point to a combination of characteristics that suggest a different temporal perspective from that described in preceding chapters. In the discussion that follows we shall take great liberties with this material, ignoring the important differences that exist among these societies in order to gain a broader perspective.

A linear notion of time's flow is partly dependent on a sense of historical change, of a past that was substantially different from the present, yet connected with it through a logical process of historical reconstruction. In societies with no written records, however, history soon melts into mythology after being retold a few times. As Baillie (1951, p. 42) pointed out, "The tribal memory seldom extends more than a few generations backwards, so that men have no knowledge of things ever having been substantially different from what they are now."

Non-literate cultures, in spite of the obvious fact that they do change through the course of history, are therefore unlikely to embrace the concept of historical change and to make of it the moving power of their development. The evidence of that change is lost in the timelessness of myth. In Lévi-Strauss' (1960) terminology, these will generally be "cold" societies that seek to deny the effects of historical change by transforming the processes of temporal succession into the sense of an unchanging, eternal, and timeless present.

When the past becomes mythological, moreover, it tends to lose its quality of distinctiveness from the present. "History and mythical reality are not 'the past' to the Trobriander," according to Dorothy Lee (1959, p. 91). "They are forever present, participating in all current being, giving meaning to all his activities and all existence." After analyzing the myths recounted by the Pawnee Indians during the nineteenth century, Wax (1962) concluded that these people viewed the past primarily as a "timeless" storehouse of tradition; it did not provide an historical record extending backward from the present and organized in terms of a calendar. "Life had a rhythm but not a progression," he wrote. "It was a series of cycles or repetitions."

Modes of time-reckoning.   Every organized social group must develop a way of coordinating collective activities, of communicating shared memories and referring to future events. As long as the society is relatively isolated and homogeneous, points of temporal reference may remain concretely wedded to recurrent natural or social phenomena, and time is expressed as the simple association of two events. Among the Nuer in central Africa, for example, communication is assured by such statements as, "I shall return at milking," or "I shall start off when the calves come home" (Evans-Pritchard, 1940, p. 102).

In most of these societies, the rhythmic alternations of natural phenomena provide the reference points, but only when important social activities are connected with them. Among the Tiv and the Trobriander, for example, the recognition of "seasons" appears to reflect primarily the return of agricultural activities rather than changes in climate or lunar divisions of time (Bohannan, 1953; Malinowski, 1935). Periods of the day during which no significant changes in collective activities occur are likely to be passed over. Discrete positions of the sun may be finely distinguished, however, during the early morning or evening hours when different activities follow one another in rapid succession (Evans-Pritchard, 1940; Hallowell, 1955). As Sorokin and Merton (1937) observed, "Time here is not continuous—its hiatus is found whenever a specific period is lacking in social interest or importance. The social life of the group is reflected in the time expressions."

The longest recognizable recurrent cycle is the year, but surprisingly, it seems to be rarely used in these societies as a measure of time. Few persons know or need to know their chronological age. The notational system makes it extremely difficult to determine precisely how much time has elapsed since an historical event occurred. Most North American Indians, for example, were reportedly able to keep a pretty close count of the passage of time within the current year, but beyond that, all chronology was indefinite (Cope, 1919).

With only concrete and salient events as points of temporal reference, a clear orientation toward the past becomes increasingly difficult the more remote the event being considered. Reference is variously made to one's physical stature at the time, for example, or to salient stages in the life cycle, or—more distantly and more vaguely—to one's genealogical ancestor living then, or to recognized historical events. Events that can no longer be connected to any known generation of human beings mark the limits of "history." At that point, "we are plunged into a bottomless mythological epoch that lacks temporal guideposts of any conventional sort" (Hallowell, 1955, p. 232).

The modes of time-reckoning used in many "traditional" societies are thus expressions of social experience, not points of division in a progressive continuum of regularly measured intervals. Generally, there appears to be no conception of clear and comparable units of time (hours, days, years, etc.) that may be added together to measure duration, or to locate different events within a single temporal order.

There is likely to be no concept of time itself as an abstract, continuous "thread," unraveling in an endless progression that links all events together while remaining independent of them. As Sorokin (1943, pp. 169–170) wrote, "None of these conceptions of time and its clock could or did emerge in the purely local societies which had no need of such a timekeeper applicable to all societies."

Time is discontinuous in these communities. Temporal divisions are used to mark certain salient social events that recur in predictable succession. Werner (1948, p. 184) suggested that such divisions are likely to be "thought of not as growing out of each other or as following each other progressively in a continuum, but as standing side by side in isolation."

The major calendar system long in use among the Balinese, for example, reflected a time conceived to be "punctual" rather than durational (Geertz, 1966). It was a device for classifying discrete, self-sufficient "days," which appeared and reappeared in endlessly repeated cycles, each a particular manifestation of the fixed order of things. Just as Marcus Simpson experienced his recollections of childhood as unconnected "images of life" lacking clear temporal reference, so the cycles recorded in the Balinese calendar, Geertz observed, "don't tell you what time it is; they tell you what kind of time it is."

The projection of repetitive cycles. With time construed in terms of recurrent events, the future no less than the past is likely to be viewed as part of the unchanging round of repetitive experiences. "The intensely crowded life of any Manus generation," according to Margaret Mead (1956, p. 68), "existed between two voids: it was tacitly assumed that the past has always been like this and the future always would be." The modes of time-reckoning we have described reflect and reinforce the projection of a "known" future.

There is a confident sense of predictability inherent in both ecological and social structural modes of temporal reference:

> Seasonal and lunar changes repeat themselves year after year, so that a Nuer standing at any point of time has conceptual knowledge of what lies before him and can predict and organize

his life accordingly. A man's structural future is likewise already fixed and ordered into different periods, so that the total changes in status a boy will undergo in his ordained passage through the social system, if he lives long enough, can be foreseen (Evans-Pritchard, 1940, pp. 94–95).

Similarly, Margaret Mead (1970) suggested that the future in such societies is largely "post-figured" for the child, as he sees others go through sequences through which he himself will later go. A close identity exists between the known past and the expected future. Much like Middy Rutherford, the child in such communities sees in his parents and grandparents an image of himself grown older, and he achieves thereby a sure sense of the future that awaits him.

It is a future constructed with a degree of detailed knowledge and conviction that is unlikely to exist in societies undergoing more rapid change or for persons who conceive of the future as a realm of new possibilities and of personal choice. Moreover, "post-figurative" cultures may persist through many years of contact with other peoples and knowledge of alternative ways of life (*Ibid.*). The feeling of difference that this knowledge entails may only reinforce the sense of one's own particular cultural identity and of its timeless validity.

The "Forthcoming" Future

Images of a future construed as a virtual certainty are of a very different sort from those constructed of imagined possibilities. Such images are likely to be fused with the present, just as much of the past in "traditional" societies also mingles with current reality through the myths that validate present experiences. The clearest analysis of this apparent fusion of present and future is provided by French anthropologist Pierre Bourdieu (1963) in his description of attitudes toward time among the Kabyle peasants in Algeria.

Much of what we in the West regard as the "future" seems to be viewed by these Algerian peasants as embodied within the concrete horizon of present experience, as a kind of "forthcoming" born of the very logic of the situation itself. In a subsistence agricultural economy, the entire cycle of production may be embraced as a single whole. The planting (in the "present") is barely dissociated from the subsequent harvest (the "forthcoming"), for all points in this recurrent cycle partake of a single context of meaning that defines the present activity. And a grain may be apprehended not only in its directly perceived properties (its color, form, texture, and so on), but also in the qualities inherent in it as potentialities (meant to be eaten, for example). Potentialities of this sort are regarded as virtually incapable of not occurring, since they are just as much a part of reality as the actual present that is directly experienced.

Beyond the limits of this "forthcoming" horizon of the present lies the purely subjective world of the "future," of imagination and desire, wherein no criteria of logic or probability need apply, and almost anything is possible. "To say 'The future belongs to God'," Bourdieu wrote in describing the peasants' view, "is above all to say that it does not and never will belong to man." Significantly, there appears to be no intermediate area between that which one can touch with one's fingers and the realm of imagination and dreams. Bourdieu's distinction between the "forthcoming" and the "future" is the difference between predictability and foreknowledge on the one hand and indeterminacy on the other.

Underlying this rigid dichotomy is a non-linear conception of the nature of time. The real world is one of rhythmic repetition and sameness. Man is an inextricable part of nature, and necessarily submissive to its recurrent cycles. Time thus construed is not the interval between events—a time of waiting and preparing, of linking present realities with future possibilities; it is instead a cyclical and repetitive "present" of which the future in many of its aspects is a known and inherent part. Events that are not known to have occurred before inhabit a world of a very different sort, one far removed psychologically from the objective realities of daily life.

The tenses of time. Metaphysical presuppositions of this sort are occasionally reflected in the grammatical forms of the language through which people construe their experiences. In his classic analysis of the Hopi language, for example, Whorf (1956, p. 57) found that it contains "no words, grammatical forms, constructions or expressions that refer directly to what we call 'time,' or to past, present, or future." Instead, there are two tenses that impose on the speaker a distinction between the same two central realms of experience to which we have referred. One tense is used to designate "objective reality," or events that have been directly experienced. It includes all past occurrences whose truth is known to the speaker as well as all present activities and their inherent future extensions. The other refers to "subjective reality," to conjectural events from the dim past of mythology and the invisibly distant, along with all the speaker's thoughts, dreams, and emotions.

For the Hopi, then, as for the Algerian peasant, there appears to be no clearly temporal future as Western societies conceive it—a future that resides in that intermediary realm of imagined events that nevertheless remain securely linked with the present, through a calculation of realistic possibilities that presupposes a linear conception of time. Instead, the "subjective reality" of such future events is informed by a very different logic from that which applies to the world of objective experience, and time does not extend into the future in the same way and along the same continuum as it does into the known past.

The metaphysics underlying our own Indo-European linguistic conventions imposes on the universe the conception of a uniformly and perpetually flowing time divided clearly into a past, present, and future as contiguous segments of a single extension, a continuum shared by all events, all subject to the same laws. "Our language codifies reality in such a way as to predispose us to view events in terms of temporality" (Lee, 1959, p. 91).

As suggested in Chapter 3, the hypothetical continuum to which we in the West attach our images of future events greatly enhances the ease with which we have come to regard that future as linearly and causally connected with present and past experiences. Would we be as readily able to do so if we spoke only the Hopi language? Whorf himself (1956, p. 154) answered as follows:

> Whether such a civilization as ours would be possible with a widely different linguistic handling of time is a large question—in our civilization, our linguistic patterns and the fitting of our behavior to the temporal order are what they are, and they are in accord.

The practice of "foresight." If it is true that members of "traditional" societies rarely construct realistic images of distant future events to motivate and guide their present behavior, how can we account for the foresight implied by the behavior patterns they exhibit? Coordinating and planning collective activities are particularly evident in peasant societies. Indeed, Zentner (1966) stated that the historical transition to stable agrarian settlements "had the effect of shifting the emphasis of men's time perspective from one in which the present is perceived in a manner which links it primarily to the past to one in which the present is linked ever more closely to the future." And Moore (1963, p. 49) dismissed as "poetic nonsense" the notion that indigenous African agriculture involves neither foresight nor planning, in view of the several-year cycle of shifting cultivation undertaken in many of the most "primitive" systems of agriculture when soil fertility has leached away.

Does the foresight implicit in such cultural patterns really reflect the imaginative anticipation of future consequences that are conceptually linked with present realities? When members of "traditional" societies are asked their reasons for current practices, they usually refer in explanation to the past rather than to the future. The systematic fidelity to a timeless tradition, Lévi-Strauss (1966, p. 236) wrote, is "attested all over the world by that endlessly repeated justification of every technique, rule, and custom in the single argument: the ancestors taught it to us." Bourdieu's (1963) analysis of future-referring behavior among the Algerian peasants led him to a similar conclusion:

> To have foresight is to follow a well marked road and not to explore new ways; it is to conform to a model transmitted by the ancestors and approved by the community. . . . Acts of foresight are much more dictated by imitation of the past and by fidelity to the values transmitted by the ancients than by forecast design of a projected future.

Barry, Child, and Bacon (1959) analyzed ethnographic reports of childrearing practices in a wide variety of cultures. They found significantly greater emphasis on training for independence and self-assertion in those societies whose economy is based on hunting and fishing. There was a far greater insistence on conformity in agricultural and farming communities, whose livelihood depends on the accumulation of food reserves for later use.

The planning implicit in agricultural activities may be sustained primarily by conformity to traditional practices, and people may participate in them without needing to construct anticipatory images of distant future events. The present, by its very nature, extends into the future through endless cycles of repetition in which later events are a part of the same actuality. The act of planting includes in its present meaning the activities of the harvest and the religious ceremonies that accompany it. And the sequence of social roles that a person will undergo is firmly "post-figured" in the roles his elders now assume, so that little of the personal future is left to be fashioned by individual initiative and present strivings.

Men and women everywhere presumably dream of living different kinds of lives, but as long as such dreams remain in a realm apart, detached from the objective realities of the present, they have little psychological status as real alternatives to current practices. On the basis of structured interviews with a sample of Blood Indians in

Canada, Spindler and Spindler (1965) concluded that "the Blood live in and think about the world as it seems to them to be. Western man lives in the world to a larger extent as he would like it to be, and thinks it should be." This of course was not always the case, and it may be useful to trace very briefly the historical forces that appear to have shaped the Western conception of linear time.

An Historical Digression

The early Greek philosophers held a view of temporal phenomena in many ways similar to those now current in the "traditional" societies as we have described them. Their general view was that they were living in a period of degeneration from the Golden Ages of Antiquity in one particular cycle of cosmic time—a phase to be followed by other identical cycles in a monotonous repetition of periodic destructions and rebirths of civilization. "History was not going anywhere," E. H. Carr (1961, p. 145) wrote in describing this conception, "and because there was no sense of the past, there was equally no sense of the future."

The early Christians abandoned the Greek theory of temporal cycles and replaced it with a view of human history as the working of divine Providence, directed toward salvation in another world. But this conception still contained little sense of development in the processes of human history here on earth. Even the Renaissance, with its confidence in human reason and its renewed secular recognition of the value of mundane existence, brought no sense of historical development. Antiquity was once again idealized, and with it came the recurrent conviction of living in another period of cultural degeneration. As long as the human condition itself failed to change, there was little basis for projecting the possibility of progressive transformations into the future.

It was not until late in the seventeenth century that a series of revolutionary events occurred that were to result in the spread of the idea of historical progress and the notion of linear time that underlies it (see Bury, 1920; Baillie, 1951). Rapid developments in the natural sciences promoted the belief in stable scientific laws and the prospect of a steady accumulation of knowledge in the future. The accelerating tempo of technological inventions, culminating in the Industrial Revolution, served to convince increasing numbers of people that the ordinary conditions of human life could be controlled by man through the power of scientific knowledge. The political revolutions in France and the United States gave the masses for the first time a sense of participating in the control of their collective destiny, and provided a blueprint for a future in which all citizens might seek personal advancement. Historians suggest that these were among the central forces that led to the future orientation, the sense of optimistic possibilities, and the view of the present as a point in historical time that links the past with a beckoning future—the notion of time's arrow that Western man has continued to take for granted.

Bell and Mau (1971) proposed a theory of social change that builds on the assumption that "a major trend of history has been an increase in the mastery of man over his natural and social environment as well as a shift in his beliefs that this is so." In "traditional" societies, they suggested, the ends of behavior are seldom questioned, and the means are often invested with unalterable religious significance. The Industrial Revolution released the means from ritualistic dogma; they were to be chosen instead according to the new criteria of efficiency and economy. In the twentieth century in Western society, the ends as well as the means of life are now increasingly freed for conscious choice. "This freedom," in the authors' view, "has been made possible for many people in advanced societies by their securing the essentials and amenities of life and by advances in philosophy and science, including social science, that have reduced people's dependence upon superstition and fatalistic conceptions of the future."

Today, revolutionary forces of a similar nature are spreading throughout the developing world. Newly independent nations have entered the political arena of the modern world. Peasant villages and nomadic tribes are being integrated into nation-states. Modern technology, schools, and factories now increasingly displace the sense of unchanging time, and the mass media bring the message of a different future to all corners of the globe. "Traditional" societies have become the concern no longer of anthropologists alone, but of historians and political scientists as well.

MODERNIZATION AND THE EVANESCENT OBSTACLES

The influence of these social transformations on the conceptions that people form of their future is the central concern of the research reported in Chapter 7. In this section, some general aspects of the modernization process are briefly explored. The time perspectives characteristic of "traditional" cultures are often thought to constitute major obstacles to social change and economic development. The available evidence, however, does not appear to support this contention.

The "Failure" of Imagination

Technical experts sent into remote areas of the developing nations are often baffled by the reluctance of the natives to accept the innovations we so "generously" offer them. The most readily available explanation is that these unfortunate people suffer from a culturally constituted inability to imagine a "better" future for themselves. In one authoritative manual prepared for change-agents, we learn that "sometimes, the traditional is so highly valued that no change is accepted as improvement" (Mead, 1955, p. 260).

Moreover, the inability to imagine or value the future may so pervade a person's life in these societies, we are told, that parents may sometimes resist any change in their child-caring practices, even if it promises to increase their child's chances of survival. The effort to reduce infant mortality allegedly meets a significant obstacle:

> It comes directly up against the belief that personal value is enhanced and increased by living; that a child's and particularly

an infant's life, therefore, is relatively without value. People who love children freely, who welcome them eagerly, may at the same time show this casual attitude toward safeguarding their lives. In the United States, prenatal as well as child care is strongly supported by the great value placed upon the future, and the equation of children with the future; but a large number of cultures value the past (*Ibid.*, pp. 204–205).

Are we really to believe that human infants are treasured by their parents only in terms of their potentialities for the future, that people who place little value on anticipating future events are therefore uninterested in protecting their children's lives? It seems far more likely that the changes in traditional child-caring practices that these medical technicians were trying to introduce were resisted primarily because their efficacy was not at all obvious to the target population.

The value of new techniques is far more likely to be perceived with regard to the treatment of disease than its prevention (Erasmus, 1961). There is real suffering and real pain to be alleviated by successful treatment, but an understanding of disease prevention can come only through the vicarious experience of formal education or from reliance on the advice of experts. The relations involved are so complex and the span between cause and effect is so extensive that the temporal connections are often impossible to perceive by observation alone. We easily forget how many of our own assumptions about the relation between present activities and future consequences are based on just such vicarious experiences, and how few of these convictions in fact derive from observations we ourselves have made.

In "traditional" societies, Daniel Lerner (1958, p. 73) suggested, people usually acquire self-conceptions that entail a minimal awareness of alternatives to current practices. They have generally achieved a clearly defined and sure sense of identity within an integrated, self-perpetuating network of ascribed roles and reciprocal obligations. Most are imbued with a deep belief in the eternal validity of their cultural ways, based in part on the suppression of memories that might disturb that sense of continuity and cultural identity. A culture of this kind, Niehoff (1966a, p. 21) wrote, "is an integrated whole, and however inefficient particular practices may be, the system does work and provides the members with a predictable future. People will not willingly give up their old ways until they are well convinced that the new ones really are improvements."

Today, the modernization process is breaking down the tight cultural integration of such societies. Newly independent nations strive to integrate their populations into political entities that will override tribal identifications. Government officials, roads, schools, and literacy invade the once isolated villages, and the mass media communicate the artifacts and models of economic development to all corners of the globe. As this process unfolds, the traditional ways can no longer securely claim the allegiance of a people who are confronted daily with the image of new alternatives, and who gradually acquire thereby a cautious appreciation for the possibility of progress.

And so, according to Geertz (1966), the "punctual" and "qualitative" calendar long in use among the Balinese is now disappearing, and "the whole conception of the relation of what is now happening to what has happened and what is going to happen is altered." Similarly, Bourdieu (1963), in a footnote to his study of conceptions of time among the Kabyle peasants, asserted that his essay should have been written in the past tense: The recent dislocations of Algerian society, the impact of colonialism, and the war with the French "have profoundly altered the system of attitudes toward time, at least in the cities and a great number of rural areas as well."

As a consequence of forces such as these, Heilbroner (1963, p. 153) wrote, "a terrible change begins to take place: the underdog wakens to his lowly position." It is with the spread of this new consciousness that *un*development becomes *under*development (Horowitz, 1972), and an indigenous political leadership strives for directed and deliberate social change in a form now impossible to subsume under the mythology of a timeless present or as simply another phase in the cyclical round of repetitive events. Lévi-Strauss' "cold" societies have become decidedly warmer, and new cultural myths are emerging to explain and promote the process of change itself.

Fundamental to this new awareness on a personal level is the phenomenon that Lerner (1958) referred to as "empathy" or "psychic mobility," the capacity to picture oneself in the other person's situation. It is not enough just to know that other peoples have different styles of living, that alternatives to current practices do in fact exist, for such knowledge may only reinforce the sense of one's own particular and timeless cultural identity. A person must also be able to picture himself participating in those alternatives and experience through his imagination the emotions generated by such newly created images. This is precisely what we saw in Chapter 1 as the essential first step in anticipating the personal future—not imagination alone, nor emotional experience by itself, but the linking of the two through the projection of an image of oneself into one's conceptions of future possibilities.

As mentioned above, it was generally believed that the "inability" of traditional peoples to imagine a different future was a major obstacle to the modernization process. The problem for economic development was thought to be one of "planting the seed of change," as Schramm (1964, p. 131) put it: "The essential point is that without raising aspirations, without stimulating people to strive for a better life and for national growth, development is unlikely ever to occur." Raising peoples' aspirations meant teaching psychic mobility, breaking through the cultural mold to change the basic attitudes of the masses; for otherwise, they would remain impervious to change, seeking only the comfort and security of the past.

It soon became clear, however, that psychic mobility is easily achieved, and "traditional" cultures intrinsically offer little resistance to expanding horizons. Writing five years later, Lerner (1963) now saw the problem as economic and structural rather than psychological: "[We] have learned that the ways of progress are hard to find, that aspirations are more easily aroused than satisfied. There is a new concern that the 1960's may witness a radical counter-formation: a revolution of rising frustrations."

In Chapter 3, we spoke briefly of the long and difficult process through which men and women in Western society gradually bring their aspirations for the future into line with reality. In "transitional" societies, when people are learning for perhaps the first time that the future need not repeat the present, that process is likely to be still more difficult, its frustrations all the more acute for the sense of betrayal that they imply. These are issues to which we shall return in Chapter 7.

The Dynamics of Fatalism

If "empathy" represents the first step in anticipating a different future, a second core ingredient is the process of temporal integration, linking those anticipations with conceptions of the past and present. If a person's new awareness and desire for change is to motivate his behavior, he must see a connection between his present circumstances and the future he imagines, finding in his conceptions of the present a basis for believing that that future may in fact be realized. Basic to any personal conception of the future as a realm of potential change, Leonard Doob (1960, p. 227) wrote,

> must be the belief that men themselves—not their ancestors, not fate, not nature, not other men—are able to control their own destinies. Here perhaps is a noteworthy candidate for one of the universal attributes, for men everywhere are not likely to seek change unless they believe that change is possible.

Thus, a second major obstacle that change-agents have discovered in the culture of "traditional" peoples is their alleged inability, presumably in spite of clear evidence to the contrary, to believe that they themselves can control the forces that will shape their future. One authoritative manual for change-agents (Mead, 1955, p. 185) stated the case as follows:

> In many parts of the world we find cultures adhering to the belief that man has no causal effect upon his own future or the future of the land; God, not man, can improve man's lot. . . . It is difficult to persuade such people to use fertilizers, or to save the best seed for planting, since man is responsible only for the performance, and the divine for the success, of the act.

Similarly, Stein (1958) attributed the lack of interest in schooling among Indian villagers in Peru to their view that "education is 'foolishness,' in the light of their self-conception as socially subordinated by destiny." And Georges Balandier (1954) called attention to the profoundly negative influence on prospects for development of cultural systems that "set little store by activities whose results are deferred for any length of time."

Yet here, too, as one examines the parameters of social life among the poor throughout the developing world, it becomes clear how much of the fatalism and resignation that they express may reflect a relatively accurate assessment of the actual limitations on action that exist in a real world of structured inequality rather than any deep-seated, culturally conditioned, psychological "incapacity." Most observers of life in traditional peasant villages, for example, have little difficulty pointing to external forces that maintain a sense of powerlessness and fatalism on the part of the persons whose experiences they record.

It is a life marked by a continuous struggle for sheer physical survival. All basic decisions affecting the village, moreover, are made in the distant cities, and the spheres in which the peasant is able to show any initiative at all are thus extremely limited. Inevitably, these villagers distrust the intentions of the national government. Yet they also have no basis for confidence in their own ability to solve the most pressing problems of their community, so they continue to expect the government to act for them. Underlying this contradiction, Friedmann (1953) suggested, "lies the deeper hiatus of [their] own personality: the inability to create a continuum between the everyday reality and the objects of [their] aspirations"—a failure of temporal integration that directly reflects the situational realities, and that cannot be attributed to cultural constructions alone.

A peasant not only has little control over the decisions handed down from outside the village, but he usually also has no idea how or why they are made. "A fatalistic attitude toward life? It is hard to imagine more favorable circumstances in which it can develop," Foster (1962, p. 48) wrote, "and hard to understand how it can be lessened until these circumstances are changed." It is surely premature to conclude that peasant fatalism, in the face of these structural realities, is simply a blind reflection of culturally conditioned personality attributes.

Still, regardless of how such beliefs are acquired, they may nevertheless affect a person's ability to take advantage of opportunities that do in fact materialize. But to what extent is this true? In the research and writings on economic development, it is increasingly recognized that fatalism in itself does not now and probably never has destroyed individual initiative or personal striving (see Niehoff, 1966b). Instead, fatalism appears to be important primarily as a rationalization for failure, a consolation when affairs are going badly. People generally resort to the belief that God alone can determine the future only after they have done everything possible themselves to avoid unpleasant experiences, to control and manipulate an unpropitious reality. "Peasant fatalism," Niehoff (*Ibid.*, p. 253) wrote, "is one of the most insignificant problems that exists in inducing change, and it becomes relevant only when innovation techniques are generally inefficient."

Under the impact of modernization, "traditional" societies inexorably undergo a process of disintegration. Images of a different future, that were once relegated to a separate realm of subjectivity, become construed as part of the objective world. The mythology of timelessness gradually fades, and change becomes the expected order of reality.

The sure sense of cultural identity that such societies once provided is now increasingly replaced by frustration and the knowledge of economic subordination. The structural changes that accompany modernization generally succeed only in transforming "traditional" peoples into "backward" peasants dependent on decisions made from afar, or into an expanding "lower class" comprised of the "culturally-deprived" and "disreputable" poor who crowd the urban slums throughout the developing world. If people confronted with such conditions tend to show little willingness to plan for the future or to delay gratification, if fatalism persists and further changes are resisted, surely the reasons are more readily found in those conditions themselves, than in any psychological inability to imagine a different future as a personal possibility.

Powerlessness and poverty exist of course in highly modernized societies as well, and individual characteristics associated with them are also frequently attributed to imputed cultural or psychological forces. The poor have remained poor, it has been said for centuries, because of their own lack of ability or initiative. In its modern form, this contention builds on the concept of culture and ascribes a predominant role to the two aspects of future orientation we encountered in the research on modernization.

THE IMPACT OF SOCIAL CLASS

Over many years of research with poor families in Mexico and Puerto Rico, Oscar Lewis (1961, 1966a, 1966b) developed his well-known conception of the "culture of poverty." It is a way of life, he suggested, that is both self-perpetuating and self-defeating, with a structure and rationale all its own, passed on from generation to generation along family lines. The culture emerged originally as an adaptation on the part of the poor to their marginal position in the class-stratified capitalistic societies of the Western world. It represented an effort to cope with the feelings of hopelessness and despair that came from realizing the improbability of achieving success in terms of the values and goals of the larger society.

What might have begun as a realistic adaptation to situational forces, however, tends to be transformed into a closed, self-perpetuating system because of its effects on the personality development of children:

> By the time slum children are age six or seven, they have usually absorbed the basic values and attitudes of their subculture and are not psychologically geared to take full advantage of changing conditions or increased opportunities which may occur in their lifetime (Lewis, 1966a, p. xlv).

Included among the traits peculiar to the subculture of poverty, we learn, is "a strong present-time orientation with relatively little ability to defer gratification and to plan for the future," as well as "a sense of resignation and fatalism."

In *The Unheavenly City,* Edward Banfield (1968) carried this argument further. The major problems in urban America, he asserted, derive from the existence of a "lower class" whose *defining* characteristic is a culturally constituted inability to imagine the future or to sacrifice present for future satisfactions. Although Lewis viewed such "cultural" traits as adaptations made by the poor to the painful realities of slum life, Banfield argued that poverty itself, in at least some of its manifestations, is the consequence of that culture, rather than its cause:

> Extreme present-orientedness, not lack of income or wealth, is the principle cause of poverty in the sense of "the culture of poverty." Most of those caught up in this culture are unable or unwilling to plan for the future, to sacrifice immediate gratifications in favor of future ones, or to accept the disciplines that are required in order to get and to spend. . . . No doubt there are also people whose present-orientedness is rationally adaptive rather than cultural, but these probably comprise only a small part of the "hard-core" poor (pp. 125-126).

How can one evaluate contentions such as these? By what criteria can one differentiate a fatalism or a "present-orientedness" based on rational assessments of situational realities from those reflecting a culturally determined inability to take account of the future? To be sure, there are personality disorders described by clinical researchers that have symptomatic traits similar to those Banfield ascribed to the "lower class culture," but such "psychopathic personalities," as we saw in Chapter 1, are relatively rare. It would seem no more than common sense to recognize that only when the objective situational forces impinging on two individuals or groups are at least roughly similar may one safely attribute differences in their behavior to internal psychological causes.

An examination of the published writings on social class differences suggests a wide range of situational realities that shape the time perspectives of persons in different positions within the hierarchy of economic and social power in American society. The conditions under which people attempt to earn a living constitute perhaps the single most decisive set of such forces.

The Temporal Implications of Work

In whatever form it takes, work usually requires the renunciation of immediate impulses, claiming time and energy that might otherwise be put to more gratifying uses. It is likely to be sustained primarily by the anticipation of future rewards. Occupational roles in industrialized societies, moreover, are saturated with temporal qualities, and not only in the importance attached to speed or efficiency, coordination and scheduling, the frequency of payments, and so on. An integral part of the well-studied role of the American business executive is "the sense of the perpetually unattained," of one goal following on the heels of another (Henry, 1949). It is this role that dominates the middle-class experience, shaping a style of life that "puts the highest possible premium on foresight and control, as against, say, retrospection and enjoyment" (Seeley et al., 1956, p. 357).

Furthermore, as Kahl (1962, p. 193) observed, "Upper-middle-class people do not have jobs, but occupy positions; they do not work, they pursue careers." The career pattern is built on "timetable norms" with a future reference inherent within them (Roth, 1963). A person's self-esteem hinges on the comparison of his own rate of progress with that of his co-workers, as he builds his personal images of future accomplishment on the experiences of his older colleagues.

In contrast, the basic fact of occupational life in the "working class" is that it generally proceeds on a flat level. As Kahl (1962, pp. 205-206) noted, "There are few differences in pay or responsibility from job to job or year to year. There's not too much point in working hard to get somewhere, for there is no place to go." If this is the case for the assembly-line worker in a factory, it is truer still for those who reach adulthood with little but unskilled hands to offer, and who live in the unstable world of poverty and failure. Menial labor in the American market rarely holds the possibility of upward movement. No amount of self-discipline or hard work is likely to transform the dishwasher in a restaurant into the chef or the manager, or to carry the janitor to a clerical job in the office building he cleans.

Moreover, the availability of such unskilled jobs varies with fluctuations in the economic system, and responds to forces well beyond the control of poor people themselves. As Allison Davis (1946) argued in a classic paper, the underprivileged worker's "goals are short term because his hold upon a job and upon clothes and upon food is short term." And jobs of this kind, even when they are available, hardly provide enough money to stay alive, let alone to support a family or to save for a child's education. When confronted with the daily pressure to satisfy fundamental physical needs, a person is forced to reduce his ambitions to the level of subsistence—to narrow, limit, and shorten his goals. Situational pressures such as these impinge in a similar manner on all who live under conditions of poverty, and class differences in outlooks on the future are inevitable consequences of these realities.

Elliot Liebow (1967) spent over a year studying a streetcorner in Washington, D.C., observing and sharing the world inhabited by a shifting collection of anchorless blacks for whom that corner was a regular gathering place. His observations of their attitudes toward work would seem to be grist for the mill of any "culture of poverty" enthusiast. The streetcorner man, Liebow reported "appears to treat the job in a cavalier fashion, working and not working as the spirit moves him, as if all that matters is the immediate satisfaction of his present appetites, the surrender to present moods, and the indulgence of whims with no thought for the cost, the consequences, the future" (p. 64).

Such words sound as if they were written by Banfield, for they suggest that these people share a self-perpetuating cultural system that renders those who are socialized within it psychologically incapable of imagining a future or of working for distant goals. The man who made these observations, however, did not perceive it that way. Liebow argued as follows:

The future orientation of the middle-class person presumes, among other things, a surplus of resources to be invested in the future and a belief that the future will be sufficiently stable both to justify his investment (money in a bank, time and effort in a job, investment of himself in marriage and family, etc.) and to permit the consumption of his investment at a time, place and manner of his own choosing and to his greater satisfaction. But the streetcorner man lives in a sea of want. He does not, as a rule, have a surplus of resources, either economic or psychological. Gratification of hunger and the desire for simple creature comforts cannot be long deferred. Neither can support for one's flagging self-esteem. Living on the edge of both economic and psychological subsistence, the streetcorner man is obliged to expend all his resources on maintaining himself from moment to moment.

As for the future, the young streetcorner man has a fairly good picture of it. . . . It is a future in which everything is uncertain except the ultimate destruction of his hopes and the eventual realization of his fears. The most he can reasonably look forward to is that these things do not come too soon. Thus, when Richard squanders a week's pay in two days it is not because, like an animal or a child, he is "present-time oriented," unaware of or unconcerned with his future. He does so precisely because he is aware of the future and the hopelessness of it all (pp. 65–66).

The Non-deferred Gratification Pattern

In our discussion of delaying capacity in Chapter 3, we saw that a child's willingness to choose a larger delayed reward over a smaller one available immediately was a function of at least two factors: his ability to endow images of short-range future events with a compelling sense of "reality," and his confidence that the promised reward would actually be delivered. The first appears to reflect a growing child's intellectual development, the second the expectations, born of past experience, that he brings into the choice situation. Some recent studies have effectively demonstrated the role of situational realities in determining the choices people make in such contexts.

Although most previous research has confirmed the expected positive relation of delaying capacity to tested intelligence and age, Bochner and David (1968) in a study of aboriginal children found the opposite to be true. School children aged seven to fifteen in a settlement in the Northern Territory of Australia were given an I.Q. test and offered the choice between one candy bar beforehand or two after the interview. As it turned out, the older children were no more likely than the younger ones to choose the delayed reward, and those claiming the immediate reward scored significantly higher in tested intelligence than did the delaying children. In the inhospitable circumstances of aboriginal life, the authors suggested, more intelligent children are likely to learn earlier that "the most adaptive thing to do is to consume within a short space of time as much as possible of what is presently available."

Another study (Rodgers, 1967) focused on three communities in the Out Island Bahamas. Two of the villages had been relocated after a hurricane, and their members were now working primarily as wage laborers in a construction industry owned by absentee Americans. The third was an isolated coastal village whose economy was based on slash and burn agriculture and subsistence fishing. Rodgers administered a questionnaire to samples of adults from the three villages, asking them to make hypothetical choices among seventeen different low-value items available immediately or comparable ones of far greater value to be had in one year's time. Respondents in the more isolated community made far more delayed choices than did those in the other two villages. The latter had abandoned their subsistence base for a money economy governed by such external forces as inflation, recession, and the decisions made by distant American managers, forces well beyond the control of the villagers themselves. The unpredictability of their economic future appeared to have eroded a willingness to postpone gratification that was still prevalent among those who had remained protected from such dubious benefits of economic development, and who were therefore able to place greater trust in the future.

It is clear that "impulse deferral" is likely to occur only when future rewards can be confidently predicted, and when people sense that the goals that they anticipate are at least to some degree under their personal control. Such conditions are rarely found among those who live in poverty. Most are likely instead to develop a fatalistic perspective on their lives, if only to protect their self-esteem in the face of repeated frustration. "If the facts of a person's past and present are extremely dismal," Goffman (1961, pp. 150–151) wrote, "then about the best he can do is to show that he is not responsible for what has become of him."

In a study of a depressed community in Nova Scotia, an interdisciplinary team of behavioral scientists (Hughes et al., 1960) observed the following pattern:

> The general feeling is that there is little to be gained in looking for work or even in thinking about future possibilities since success or failure is governed only by luck (which can operate

without any effort on the individual's part) or "pull" (which
they know they do not have). So on the whole there is very
little anticipation of or planning for future events (p. 303).

Srole (1956) operationalized Durkheim's concept of "anomie" into
an attitude scale designed to assess (basically) the degree to which
an individual feels hopeless and discouraged in his outlook on the
future. In interviews with 701 men aged twenty-one and over, Meier
and Bell (1959) found that the best predictor of such attitudes was
social class: The lower the respondent's socio-economic status, the
more likely he was to agree that "It's hardly fair to bring children
into the world, the way things look for the future," or that "Nowadays,
a person has to live pretty much for today and let tomorrow take
care of itself." Similar findings that further demonstrated the relationship between social class and perspectives such as these were
reported by Rosen (1956) and Strodtbeck (1958) using high school
students as respondents.

Another scale was developed by Rotter (1965) to measure a person's belief that events in his life are generally contingent on his own
behavior (internal control), rather than on external forces such as luck,
fate, or the control of powerful others, or viewed as basically unpredictable because of the great complexity of the forces surrounding
him (external control). This distinction, among others, will be central
to the growing conflict between the two people on whom the life study
in Chapter 8 is based. To demonstrate the validity of the scale, Rotter
cited several studies that revealed a significant positive correlation
between socio-economic status and the belief in internal control.

One of them (Battle and Rotter, 1963) was particularly suggestive.
Among a group of sixth and eighth graders, lower-class black children
of high tested intelligence showed a significantly greater belief in
external control than did middle-class whites of low I.Q. Moreover,
the less intelligent lower-class black children were more likely to
believe in internal control than were the high I.Q. lower-class blacks.
The similarity between these findings and the investigations of

Bochner and David suggests that here, too, more intelligent lower-class children may be quicker to realize that their lives are in fact largely under the control of external forces.

If lower-class persons generally tend to see the future as relatively unpredictable or subject to forces over which they have little control, they should also be far less likely than those in the middle class to sacrifice present for future rewards. Indeed, the presumed absence of a "deferred gratification pattern" has become a major conceptual tool in the analysis of lower-class behavior patterns. The class-related characteristics that have been taken to reflect an inability on the part of the poor to postpone impulse gratification cover an extraordinarily wide range. They include, for example, more frequent pre-marital sex, lower aspiration levels and educational attainment, the lack of financial savings, and a greater propensity for physical violence (see Schneider and Lysgaard, 1953).

And yet, as Allen (1970) indicated, surprisingly little direct empirical research—in contrast to inferences drawn from pre-existing surveys—has seriously endeavored to test the notion that the poor are in fact more "self-indulgent." Undoubtedly, real class differences do exist in attitudes and behavior patterns, but such differences should be cautiously interpreted before concluding that they represent deep-seated differences in personality traits. They may be reflecting the social reinforcements and situational constraints that operate in the present to shape a person's behavioral choices.

Social class and social norms. Observers of working-class communities in the United States have often noted that the values by which their members relate to one another appear to be somewhat different from those that typically govern middle-class behavior. Broadly conceived, the distinction generally relates to what Albert Cohen (1955) labeled the "ethic of reciprocity" in contrast to the "ethic of individual responsibility." The working-class person is often deeply embedded in a network of friends and relatives, within a system of mutual rights and obligations. In times of adversity, he will turn to them for aid and draw on their resources, and he remains ready to share his own resources with them when the need arises. The middle-class ethic of individual responsibility, on the other hand, minimizes the obligation to share with others, particularly when that obligation might interfere with the pursuit of one's own achievement goals.

Cohen pointed out that the ethic of reciprocity and the desire to have a good time here and now in the company of one's friends do not by definition preclude the desire to help oneself and to provide for the future. They do, however, make acting on that desire somewhat less profitable, for whatever material success one might achieve by one's own efforts is likely to be quickly dissipated in the obligation to share it with others. Only by making a prior break from his family and peer-group relationships can the working-class youth safely pursue long-range personal goals of occupational or individual development (Gans, 1962). In his classic study of an Italian slum community in Boston, William F. Whyte (1943, p. 107) noted that "both the college boy and the corner boy want to get ahead," but "the corner boy is tied to his group by a network of reciprocal obligations from which he is either unwilling or unable to break away." And Douvan and Adelson (1958), in their national survey of American high school students, found that downwardly mobile adolescents were significantly more likely to mention instances of peer acceptance as major sources of personal pride than were boys of comparable social-class backgrounds who aspired to occupations above those of their fathers.

Further suggestive evidence in this connection comes from extensive research in Israel comparing the personality traits of children brought up in the Kibbutz with those of family-reared children (Rabin, 1965). The non-Kibbutz children, both fourth graders and high school seniors, were significantly more likely to give answers on projective tests that indicated long-range personal goals and more extensive perspectives on the future. The Kibbutz children, on the other hand, showed a far greater concern with long-range social goals, for their involvement in the collective structure of the Kibbutz made it unrealistic to develop personal aspirations for the distant future.

The available data are consistent with the notion that an extended perspective on the personal future is facilitated by a clear sense of self-reliance and individualism (or isolation?), which encourages a person to view his prospective life span as his own possession, dependent on his personal initiative. To the extent that it permeates the working class, the "ethic of reciprocity" may be partly responsible for social class differences in orientations toward the personal future. It does not, however, imply a greater tendency to indulge immediate impulses. The working-class youth who shares his material possessions with his friends is, after all, renouncing the personal gratifications that he might otherwise have been able to enjoy. Granted, he gains other rewards in the exchange, but it is questionable whether the bargain is very different from the calculations underlying middle-class patterns of "deferred gratification."

The calculus of satisfactions.   Middle-class adolescents are generally far more likely than those in the working class to stay in school or go on to college. To what extent may we safely infer that this reflects a greater ability on their part to sacrifice present for future rewards? Are college-going youth really forced, as a general rule, to give up spending-money, good times, interpersonal relationships, or travel in order to pursue their educational goals? Are they really unable to find in the college experience itself present satisfactions that outweigh those they might have obtained by leaving school and going immediately to work?

Beilin (1956) interviewed working-class high school senior boys, roughly half of whom intended to go to college. The students who were going on to college gave no indication at all that they saw themselves as delaying gratification. Rather, they viewed the college experience as a means of achieving particular satisfactions that were more significant to them personally than those desired by the non-college-going youth. For both groups, moreover, these were predominantly short-range objectives. As the author concluded, "It would appear that postponing is a phenomenon the observer introduces to explain apparent differences in behavior although the actors themselves do not perceive that they are behaving in this manner." If this is generally the case among working-class adolescents, it is even more likely to be true of upper-status youth who, as Kahl (1962, p. 287) discovered, "aimed toward college as a matter of course, because everybody in their group did so."

Similar questions need to be raised regarding the inferences that are usually drawn from Kinsey's (1948) well-known disclosure that male respondents with a grade school education report an average rate of pre-marital intercourse approximately seven times greater than that of the college population. If this reflects a class-linked inability to delay gratification, why were not similar class differences found for females? And what about the finding that middle-class males engage in far more petting and masturbation? Such class differences would appear to be related more to the manner of achieving sexual gratification than to the amount of gratification actually achieved or deferred (see Miller, Riessman, and Seagull, 1965).

In spite of its apparent reasonableness, the assumption of a clear relationship between social class and the willingness to delay gratification finds little direct empirical support. In addition to Beilin's research, two other well-conducted studies (Seagull, 1964; Shybut, 1963) sought to investigate this relationship using realistic behavioral measures of delaying capacity. In neither case were social-class differences discovered.

Within the broad spectrum of the American stratification system (ranging from the stable working-class poor with steady blue-collar jobs to upper-middle-class professionals), individuals clearly differ in the value orientations that shape their interactions with others and influence the satisfactions they seek. But there is little basis for concluding that social-class position within this status hierarchy is associated in any consistent manner with a person's willingness to postpone his gratifications.

As each of us becomes acutely aware of the irreversibility and finitude of our lives, we are faced at every turn with the same broad alternative: Shall we choose the immediate reward, or work instead for the remoter but greater one? As MacIver (1962) observed, this is indeed one of life's major dilemmas, and most of us, whatever our social status, come to see the value in both alternatives. Even under the most optimistic circumstances, future gratifications are not always to be trusted. There is a line from Dryden that states: "Present joys are more to flesh and blood/Than a dull prospect of a distant good." And however we may resolve the dilemma at any given moment, its resolution "may represent the triumph, sometimes by a narrow margin, of one value system over a weaker but nonetheless serious competitor" (Cohen, 1955, p. 109).

If it is difficult to find evidence of greater present orientation among stable working-class people in American society, restricted time perspectives may nevertheless prevail among the "unstable" poor in the urban slums and in the poverty "pockets" of rural America. It is within these areas, and among those who have yet to share in the growing affluence of this society, that objective realities are likely to create the conditions for a genuine and pervasive present orientation.

The Lower Class Revisted

The daily pressure of satisfying basic physical needs, the instability of income, the improbability of moving out of menial jobs, the realistically bleak expectations, and the fatalistic perspectives that offer the sole protection for battered self-esteem—these are the conditions of poverty, and until they change the people caught within them are unlikely to develop extended future orientations or to sacrifice present satisfactions for imagined future ones. These realities are finally acknowledged even by those who have most emphasized the psychological dimension in accounting for the behavior patterns generally found among the poor.

Oscar Lewis (1969) recently contradicted the implications of his earlier formulation:

> The crucial question from both the scientific and the political point of view is: How much weight is to be given to the internal,

self-perpetuating factors in the subculture of poverty as compared to the external, societal factors? My own position is that
in the long run the self-perpetuating factors are relatively minor
and unimportant as compared to the basic structure of the larger
society.

Similarly, Banfield (1968, p. 223) acknowledged that his own analysis
of "lower class culture" must be regarded as highly tentative. "It may
turn out that the lower class that has figured so largely in this book,"
he wrote, "does not exist (that whatever present-orientedness exists
is neither cultural in origin nor cognitive in nature) or that it exists
among so few persons as to be inconsequential." The great bulk of
research on social class differences in the United States would suggest
that this is indeed the case.

Perspectives on the personal future are profoundly shaped by the
social and cultural context in which a person is embedded. The experiences that underlie one's conceptions of the present and its projec-
tions into the future derive from that context. It is therefore to these
cultural and social realities that one must look first for explanations
of individual differences in outlooks on the future.

The conditions of poverty and deprivation are primarily responsible for the feelings of powerlessness and the restricted present
orientations prevalent among the "lower classes" in American society.
Such conditions may have somewhat different effects, however, in the
developing world, when the processes of modernization first release
the future from its "post-figurative" predictability. The empathy that
frees imagination from the bounds of the past and the present provides
little basis in itself for the development of that constrained imagination that distinguishes hopes from expectations in the images that a
person forms of the future. We still have much to learn about the
particular social changes that are likely to be most important in this
connection, and about the processes through which they are translated
into individual differences in outlooks on the personal future. These
issues are explored in the next chapter.

7
Modernization
and the Transforming Future:
Some Psychological Consequences
of Social Change in Tunisia

> Men resemble their times more than
> they do their fathers.
>
> Arab Proverb

We have spoken in general terms about the likely
influence of modernization on the personal future as individuals in
developing societies conceive it. In this chapter, we present some
preliminary findings of a research project conducted in Tunisia
(Klineberg, 1971, 1972). Our central concern is with the impact of
social change on the images of the personal future held by adolescents
in a poor but stable section of the inner city of Tunis during the
early months of 1970.

THE RESEARCH SETTING

Tunisia is an Islamic country of some 5 million people in an area
roughly the size of New York State, situated between Algeria and
Libya on the North African coast. Ruled as a French protectorate

since 1881, it became an independent nation on March 20th, 1956, less than fourteen years before our research began.

During that long colonial period, profound social changes were imposed on the traditional order of Tunisian life. The French established their organizational control through a network of local administrators in all regions of the country. Agricultural lands were seized, and the most advanced methods of farming were introduced for crops designed primarily to be exported to France. Rapid population growth in the rural areas coincided with declining opportunities for work on the farm. With this increasing threat to the traditional subsistence economy, rural people flocked to the cities, where unemployment was reaching alarming proportions. The Tunisian economy remained dependent on fluctuations in the world markets, and few positive measures were taken by the French in an effort to benefit the population as a whole.

By 1956, fully one-fourth of the labor force was unemployed or underemployed. Only 9 per cent of the population was engaged in the industrial sector, and 274 of the 304 largest firms were run by Europeans. Three out of every four Tunisians still struggled for subsistence in the rural areas. On the eve of independence, only 27 per cent of all primary school age children were actually in school; only one Tunisian child out of thirty was attending secondary school.

Throughout the colonial period, the modern life styles represented by the French were generally resisted by the masses as a threat to their Tunisian identity. Most of the population, in the cities as well as in more isolated rural areas, remained deeply integrated into a sacred social order, their interpersonal behavior regulated by detailed rules whose validity was unquestioned (see Camilleri, 1970).

It was a social order based on the patriarchal family and the absolute authority of the father. It was he who planned the meals, who determined what clothes his wife would buy, what man his daughter would marry. His voice was sacred and powerful; obedience to him was commensurate with obedience to Allah Himself (see Demeerseman, 1967). Women and girls above the age of ten were usually kept within the house; if they did go out, they were heavily veiled. They were forbidden any form of social life or job opportunity, except as artisans in their homes. Everywhere, they were considered inferior, to be kept in fear and awe of their masters.

Hardly a dent was made in this pervasive family structure during the colonial era. As Micaud (1964, p. 146) pointed out, "The prestige of the Great Mosque, strengthened by the French authorities, the forces of tradition, and the caution that the Neo-Destour[1] had to exercise in the domain of religion, prevented any significant change of customs." With the advent of independence, however, the life styles typical of Western society now became national goals, to be spread throughout the country in an effort to forge a new social identity as a modern nation. During the past decade and a half, "a massive effort has been made to transform attitudes and values and to reduce psychological and social obstacles to progress" (Ibid., p. 131).

In the years immediately following independence, the Tunisian economy remained stagnant and even deteriorated. Capital investments rapidly declined with the massive outflow of European funds. In the climate created by the Algerian war, the French cut off economic aid, and only massive and continuing injections of foreign capital, especially from the United States, prevented a catastrophic fall in standards of living. Efforts to deal with such basic economic problems were deferred, and energies were focused instead on creating a spirit of national unity, on consolidating the new state through a shared commitment to social progress, and on developing a skilled labor force to replace the departing Europeans. "The modern nation-state," as Micaud (1964, pp. 136–137) described the grand strategy, "was to be built on civic virtues—and only afterward on economic accomplishments."

Within a year of independence, broad legal reforms had replaced Koranic law, bringing all Tunisians under the jurisdiction of com-

[1]The political party that under the leadership of Habib Bourguiba successfully mobilized the revolutionary potential of Tunisian society and led the fight for independence.

mon-law courts. Polygamy was prohibited. Marriage, that for centuries was an agreement between two families, became a contract between two persons whose explicit consent was required by law. The minimum age for such consent was raised, and new divorce laws were now based on the principle of complete equality between the sexes.

The call to national unity and the commitment to modernity soon spread throughout the country through skilled and intensive use of all the media of mass communication. President Bourguiba's message was direct, forceful, and easily comprehended. His frequent speeches were eagerly awaited, carried by newspaper and television and heard over the blaring radios in the cafes and shops. Party cells were active in every town and village, organizing cultural activities and national celebrations and amplifying the message as if they were continuously waging an electoral campaign (Moore, 1964).

The most far-reaching reform of all was the dramatic expansion of access to formal education for girls as well as boys during the decade and a half since independence. In a speech at Sadiqui College in 1958 (cited in Brown, 1965), Bourguiba stated: "When we were in the opposition and Tunisia belonged to others, not to us, we planned and resolved that when our country was independent and the state apparatus in our hands we must treat first the problem of education."

Tunisia's investment in public education in proportion to national income is among the highest of any country in the world. Enrollment in primary schools has shown a fourfold increase since independence, and over five times as many Tunisian children are now in secondary schools. Over 70 per cent of all primary school age children throughout Tunisia are in school this year, and that figure approaches 95 per cent in the largest cities.

Thus, the signs of change are everywhere today, and one senses that the traditional culture, once accepted unquestioningly as the natural order of human existence, has become for an increasingly large proportion of Tunisians a way of life to be examined and weighed against a new awareness of alternative possibilities. The future is no longer expected to follow the models of the past, and the ancient certainties are giving way to a new cultural ambiguity, as different relationships between men and women, parents and children, citizen and state, now gradually emerge.

It was into this setting that we arrived in the fall of 1969 to conduct a study of the self-conceptions and attitudes of adolescents and their parents living in *Bab Souika*, the most self-contained and most self-consciously traditional of the sectors comprising the inner city of Tunis. We were extremely fortunate in being able to collaborate closely during the interviewing phase of this research with Professor Taoufik Rabah, now Director of Research at the Institute of Business Management, University of Tunis. It was he who took major responsibility for translating the questions we devised into colloquial Arabic and into a form appropriate to the Tunisian cultural experience.

There were some 10,500 families living in *Bab Souika* at the time of the 1966 citywide census. We examined 500 of these dossiers, selected at random, and all families with one or more children aged thirteen to nineteen were included in our target sample. Meanwhile, we had designed four different questionnaires, varying slightly according to whether we would be speaking with parents or adolescents and with students or those who were no longer in school. Conducted in colloquial Arabic, the interviews sought to measure a wide range of attitudes, values, experiences, and knowledge applicable to the Tunisian context. We had thoroughly pre-tested and revised our questions in the light of extensive pilot studies conducted earlier in the year.

Our interviewers, twenty-nine men and eight women, were Tunisian university students who were trained in interview methods and had participated in the earlier pilot research. During the three weeks of interviewing (in February and March of 1970), they worked in teams, talking with all respondents in a family simultaneously, often in the four corners of a single room. The interviews lasted between one and three hours, averaging slightly over an hour and a half with each respondent.

During the course of the interviewing, we discovered that fifty-five of the 196 families on our original list had moved and thirteen addresses had disappeared in the urban renewal since 1966. These we replaced where possible with comparable households living at the same or immediately adjacent address. By the end of that three-week period, we emerged with 502 completed interviews from 161 different families.

All adolescents in each family had been questioned as well as a randomly selected 70 per cent of their parents. Eighteen of the household heads told us in the course of the interviews that they were an older sibling, cousin, or grandparent of the adolescent in the family. Since we wanted to maintain a clear and consistent definition of the "parental generation," these respondents were excluded from later analyses of the data. The final sample thus consisted of 484 people: 226 adults (109 fathers, 117 mothers), all biological or adoptive parents of 258 adolescents (138 boys, 120 girls).

FAMILY BACKGROUNDS AND INTERPERSONAL RELATIONSHIPS

We had chosen a traditional area of the inner city because we were interested primarily in the impact of social change on people who did nothing themselves to search out that change. Well over half the parents in our sample were born in Tunis, as were 85 per cent of their adolescent children. They were simply there, when Tunisia won its independence. Soon, going to school became the expected occupation of children. Bourguiba's speeches, the legal reforms, and the collective power of the mass media were calling with increasing insistence for new perspectives to replace the fatalism and partriarchy of the traditional culture. We wanted to know how these particular people were responding to the social changes with which they were now confronted.

The contrasts between parents and children suggest that we do indeed have a sample representative of the "transitional" generation in Tunisian society. Forty-three per cent of the fathers and 87 per cent of the mothers had no formal education whatsoever. All of their children have gone to school. Indeed, of the 258 adolescents included in our sample, 179, or almost 70 per cent, were still in school at the time of our interviews. Of those who had left school, over 83 per cent had attended at least five years, and 17 per cent had had some secondary education. Two-thirds of the adolescents who were still in school were now at the *Lycée,* and the rest without exception expected to be there within the next few years. Moreover, although 78 per cent of the mothers in our sample said they always wear the veil when they go out, 76 per cent reported that their daughters never do.

The patterns of attitude and behavior among the parents generally conform to the traditional patriarchal structure. The women remain relatively isolated from the external currents of social change, and are largely confined within the home. One of the questions we asked, for example, was, "Whom do you usually confide in with regard to your personal problems?" Less than 17 per cent of the mothers named a person who was not a relative, whereas almost 40 per cent of their husbands said they confided in friends outside the extended family. Less than 17 per cent of the mothers could name the president of the United States; only 4 per cent knew the name of U Thant, then Secretary General of the U. N. The comparable figures for the fathers were 66 per cent and 43 per cent, respectively.

The traditional segregation of husbands and wives and the relative lack of communication between them was revealed in questions pertaining to family planning. We asked the parents in our sample to tell us how many children they personally would recommend to a young couple seeking their advice today, and what contraceptive method, if any, they would suggest. We also asked them how many children they thought their husband or wife would recommend and the contraceptive he or she would advise the young couple to use. Our data, based on 106 parental couples, suggest that husbands and wives have little knowledge of each other's attitudes in this important area.

Women think that their husbands want far more children than they really do; men think that their wives want fewer. Both significantly underestimate the degree of knowledge that the other has of contraceptive methods. As one of the women put it when asked what method her husband would recommend, "He doesn't know anything. If he knew what to do, we wouldn't have had so many children!" Indeed, such faulty communication may be partly responsible for the fact that well over half the adolescents in our sample report that they have six or more siblings, and less than 9 per cent come from homes with fewer than four children.

The woman's realm remains the home, the man's the outside world. As the families within our sample increase in socio-economic status (as measured by the father's education and occupational level and the family's household possessions), the women develop more "modern" perspectives, but almost exclusively in their role as mothers. The women of higher socio-economic status are significantly more likely than those of the same age but of lower social status to say that they are raising their children differently from the way they themselves were raised—more in keeping with the newly emerging demands of modern life. They would also allow their sons and daughters to travel abroad at an earlier age, and they express a deeper belief in their children's freedom to choose their own spouses without parental interference and to seek personal success wherever it might lead them.

Among their husbands, increasing socio-economic status brings changes of a contrasting sort. The men in our sample who have achieved higher educational and occupational levels are no different from those of lower social status in their perspectives on their roles as fathers. Instead, they evidence a far greater interest in and knowledge of current events, and they are more aware of the changes that have occurred in the life of their neighborhood. They also view both themselves and their wives as far more "open to the changes of modern life," and they expect their children to enter significantly more prestigious occupations. None of these perspectives is related to socio-economic variables among their wives.

Increasing social status is accompanied by greater attitudinal differences between husband and wife, as each moves psychologically in contrasting directions with increasing exposure to modernizing influences. Despite the unprecedented legal and social reforms undertaken in Tunisia since independence, the traditional distinction between home and society and the specialization of functions between the sexes continue to affect the lives of Tunisian adults, mediating their responses to social change and keeping men and women in separate social orbits that rarely appear to intersect.

In their attitudes toward their children, the parents generally maintain these traditional sex-role distinctions. Whereas 35 per cent of the fathers, for example, said that it would be acceptable to them if their son married a foreigner, only 6 per cent would offer that same freedom to their daughters. Asked at what age they would allow their daughters to travel abroad, 27 per cent said "Never!" Less than 4 per cent would restrict their sons in the same way.

THE DETERMINANTS OF ADOLESCENT ATTITUDES

When we explore the impact of sex roles among the adolescents themselves, a very different pattern emerges. Controlling for amount of schooling, age, and socio-economic status, the boys in our sample continue to show a greater knowledge of current events and a deeper sense of personal control over their lives. But the girls as a group are now equally committed to their independence from traditional family obligations, their occupational aspirations are equally high, and they are more optimistic about the changes they perceive in their society. Above all, these young women are far more committed to sexual equality: They expect to have significantly fewer children than the boys do, and they are more determined to raise their own children differently from the way they themselves were raised. Although the boys follow their parents in making a clear distinction between behaviors appropriate for sons and daughters, the girls plan to give their own daughters virtually the same freedoms as they give their sons.

These findings suggest that the traditional sex-role distinctions that are still clearly apparent in the parental generation are changing for adolescents today, and these young people are now searching for new definitions. The evidence points to the development in Tunisian women of perspectives and self-conceptions that indicate little willingness on their part to accept the traditional roles of wife and mother as their society has long defined them. The data also suggest that they may have considerable difficulty finding future husbands who will support them in the new identities that they have developed.

In comparison with their own parents, sons and daughters alike display a greater knowledge of current events, a keener sensitivity to the changes that have occurred in their neighborhood, and a deeper belief in their personal control over the forces that will shape their lives. In addition, they are far more in favor of coeducation, they would allow both sons and daughters to date at a much earlier age, and they want fewer children than their parents would recommend. What forces account for such generational transformations? Is it the new experience of schooling alone, or are parents equally important in encouraging the development of these more "modern" perspectives in their children?

The Relative Contribution of Parents and Schools

Only recently have systematic research efforts been made to explore such issues (see, for example, Armer and Youtz, 1971; Doob, 1960; Inkeles, 1966, 1969; Kahl, 1968; Lerner, 1958; Schnaiberg, 1970). Most point to formal education as the pre-eminent influence. Indeed, as Inkeles (1966) observed, "Almost all serious investigations of the question have shown the individual's degree of modernity to rise with increases in the amount of education he has received." The exact role of education, however, is a complex matter, and few empirical efforts have been devoted to unraveling this dynamic of social and psychological change.

Formal education is acquired early in life when a child's experiences are still largely under the control of his parents. Parents who are themselves more "modern" may be more likely to encourage their children to continue their schooling, while also attempting to pass on to them values similar to their own. The association between education and "attitudinal modernity" among adolescents may be primarily a reflection of parental values operating as the central determinants of each. Controls for standard background variables are insufficient safeguards in this connection, for parents with similar degrees of education or occupational attainment may differ among themselves in the extent to which they encourage their children to develop more "modern" attitudes. The present research, by including interviews with parents along with their adolescent children, makes it possible to determine which sets of forces (parents or schools) appear to be the primary determinants of the modernity of attitudes among the adolescents in our sample.

It is commonly assumed that schooling introduces a radical discontinuity between the generations in developing societies, as youth are resocialized through the educational process for adult roles very different from those anticipated in their earlier upbringing within the "traditional" home (McQueen, 1968). This assumption seems highly questionable. Tunisian parents may well have had far less schooling than their children now obtain, and they may still be working within the traditional sectors of the economy; but they may be no more immune than their children to the forces of social change and the impact of the mass media. Previous research suggests that parents themselves are generally among the prime mediators of social change, aware of the new demands that the future will bring, rearing their children in ways that often differ substantially from the way they themselves were raised (Aberle and Naegele, 1952; Inkeles, 1955). Little effort has been made, however, to investigate this process in the developing societies.

Our study also differs from most previous research in this area by giving as much attention to girls and mothers as to boys and fathers. The usual assumption appears to be that the men in developing societies are the ones most directly exposed to the influences of modernization through their activities outside the home, and that it is on their changing attitudes and values rather than on those of women that successful development is likely to depend. The major studies in this area have focused almost exclusively on male respondents. Women are typically included only as objects of questions addressed to men, as the "modernism" of the latter is assessed in terms of the degree to which they would grant a modicum of dignity and respect to the former. As a result, we know little about the role of sex differences in responses to rapid social change, and still less about the forces that contribute to "attitudinal modernity" among young women in developing societies where their traditionally subordinated status, as we have seen in the case of Tunisia, is now undergoing a slow but inexorable transformation.

Tunisian parents, however, generally continue to reflect the traditional sex-role distinctions in their attitudes toward their children, and they are less likely to encourage the development of more "modern" perspectives in their daughters than in their sons. Schools may therefore play a more uniquely powerful modernizing role for young women in such developing societies.

This is precisely what we found in an analysis of the relative contribution of parents and schooling to the "modernity" of the adolescents in our sample (Klineberg, 1972). We devised two independent measures of "parental modernism" based on an analysis of the intercorrelations among the attitudes the parents expressed. The first reflected their sensitivity to changes in the life of their neighborhood, their knowledge of current events and, above all, their efforts to raise their own children differently from the way they themselves were raised. The second measured primarily the parents' belief that a man is responsible for his successes and failures, and their conviction that each person, their children included, should search for his own fulfillment free of traditional family obligations. Low scorers on this dimension were likely to agree, for example, that "When a man is born, the success he will have in life is written in advance," or "In looking for work, it's better to find a job near one's parents, even if it means losing a better opportunity elsewhere." Do parents who are themselves more open to change and "modern" in their perspectives on life pass on such attitudes to their children, regardless of the amount of schooling the latter have obtained?

For the boys in our sample, the answer appears to be "yes." The data reveal, for example, that the fathers' attitudes rather than schooling are responsible for differences among the boys in their belief in the efficacy of planning for the future, in their childrearing intentions, in their feelings about the importance of religion, and in the way they spend their leisure time. Among the boys who are still in school, moreover, how far they expect to be able to continue their education appears to be a function not of the amount of schooling they have already received, but exclusively and significantly of their fathers' attitudes, particularly their fathers' belief that success is a matter of personal effort and individual initiative.

The influence of schools on the adolescents' attitudes is by no means negligible, although it does appear to be relatively circumscribed. For the boys in our sample, it is the amount of schooling they received, rather than parental modernism, age, or socio-economic status, that accounts for differences among them in their knowledge of current events, their attitudes toward coeducation, their belief that they are better off than their fathers were at their age, and their occupational aspirations.

The data in general are clear in their suggestion that fathers continue decisively to influence their sons, even as the sons acquire far more schooling than the fathers themselves obtained. There is little suggestion here of a radical discontinuity between the generations, occasioned by the spread of formal education. Indeed, the boys generally appear to be at least as much influenced toward more "modern" perspectives by their fathers' attitudes as they are by the experience of schooling itself.

For the girls in our sample, the situation seems to be strikingly different. It is the amount of schooling they have received rather than their parents' attitudes that significantly influences their intention to raise their own daughters differently from the way they themselves were raised, their belief that children have the right to disobey their parents, and the sophistication they reveal in recalling a recent event in the news. Educational aspirations, which in boys are related exclusively to their fathers' attitudes, are influenced only by schooling and age in the girls who are still in school.

Only among the girls, moreover, do we find that simply remaining in school significantly affects the attitudes these adolescents hold. The girls who have left school have far lower occupational aspirations, less knowledge of current events, and a more fatalistic outlook on the future than do those girls of comparable educational attainment, age, socio-economic status, and parental values, who have managed to stay in school. Among the boys, on the other hand, those who have left school differ only in their occupational aspirations from their peers who have remained.

Formal education is clearly of particular importance as a modernizing influence for the girls in our sample. Parents seem to offer their daughters relatively little encouragement as they develop more "modern" attitudes and aspirations. When a boy leaves school, at least in the Tunisian community we studied, he enters the established world of masculine prerogatives. He remains in the company of his friends, equally in contact with the changes that surround him, with the mass media and the new perspectives that they reinforce. When a young woman leaves the school system, however, a patriarchal system awaits her return. Her leisure activities are now confined to the home and the company of her parents. When we asked the girls in our sample whether they had a group of friends with whom they spent their leisure time, almost 40 per cent of those who were no longer in school said "No," compared with less than 17 per cent of the girls who had managed to stay in school.

The social changes now transforming Tunisian society have nevertheless reached these particular young women. In comparison with their mothers, the girls who have left school are far less fatalistic about the future. They also want fewer children than their mothers would recommend, and they see themselves as much more "open to the changes of modern life" than do their mothers in perceiving them. As Micaud (1964, pp. 148–149) observed, "Upon leaving high school, a girl returns to the veil, her home, and her father's authority. But chances are that she will have a voice in the choice of her husband and later in the raising of her children." Indeed, the girls in the sample who are no longer in school are just as insistent as the school attenders on their right to choose their future husbands, and they are even more likely to say that they plan to raise their own children differently from the way they themselves were raised.

The Dynamics of Occupational Aspirations

Among the school attenders, 48 per cent of the girls and 68 per cent of the boys expect to acquire university training beyond the high school level. When asked what job they thought they would have at the age of twenty-five, over 61 per cent of the boys and 52 per cent of the girls said they would be professionals, such as doctors, lawyers, or college professors; another 30 per cent saw themselves in white-collar jobs as school teachers, policemen, secretaries, or nurses. When asked to list all the events that they thought would happen to them in the future, only 17 per cent of the girls spoke of marriage or children, whereas more than 67 per cent of their total answers reflected educational attainments or occupational success. Such images of the personal future are particularly striking when we consider that all but 17 per cent of their own mothers are illiterate, and all but 8 per cent are unemployed.

There is little doubt that these Tunisian adolescents have defined themselves and their futures in ways that contrast markedly with traditional expectations and parental models. Although 40 per cent of their fathers hold unskilled or semi-skilled blue-collar jobs, not a single one of the boys in our sample of school attenders expects to follow them into occupations at that level. Although only 10 per cent of the fathers are in high white-collar or professional positions, over 60 per cent of the sons expect to enter these kinds of occupations.

For both sexes, occupational aspirations are primarily influenced by the amount of education these adolescents have received. Only 36 per cent of the boys and 29 per cent of the girls still in elementary school expect to enter high white-collar or professional positions, but the proportions rise consistently with grade level until they reach 79 per cent and 83 per cent, respectively, among those in the later years of secondary schooling. This close relationship between education and aspiration persists at statistically significant levels even with rigorous controls for differences in age, socio-economic status, and knowledge of current events.

"Empathy" has clearly taken hold, and these young people now picture themselves in social roles and occupational positions that would scarcely have been imaginable to their parents during their own adolescence. But the aspirations they express reflect little acknowledgement of the realities of Tunisian opportunities. The one university in the country can enroll fewer than 10 per cent of the students now completing high school throughout the nation; yet we have seen that well over half of the adolescents in our sample of school attenders expect to acquire university training. When asked if there were any obstacles that might conceivably prevent them from attaining the occupations they envisioned, 47 per cent could think of none at all, and only 25 per cent mentioned the possibility that they might be unable to fulfill their educational plans.

The future has been released from its "post-figurative" predictability (Mead, 1970). The sense of new possibilities that this engenders has severed the close identity between past or present realities and anticipations of the future. But empathy alone provides little basis, as we suggested in Chapter 6, for the development of that constrained imagination that distinguishes hopes from realistic expectations.

Our measure of occupational aspirations was the job these adolescents thought that they would have at the age of twenty-five. Later in the interviews, we asked them the following question: "If you could have *any job at all*, what job would you choose?" Neither boys nor girls made any consistent distinction in their answers to these two questions. If anything, there was a slight decrease in the mean prestige level of the occupations that they would ideally choose in comparison with those that they "expected" to have. The parents reflected a similar pattern: Both fathers and mothers "expected" their children to have jobs at the age of twenty-five that were actually higher in levels of income and prestige than the ones these parents could ideally imagine for themselves!

Rubin and Zavalloni (1969) collected autobiographies of the future from secondary school students in Trinidad and Tobago. These adolescents also expressed unrealistic fantasies of future fulfillment, reflecting in their essays a strong drive for recognition and great accomplishment, and an overwhelming faith in the "ethos of mobility." Similar findings were reported by Foster (1965) in a survey of high school students in Ghana.

As long as adolescents remain in school, it seems, the future is subjectively far away. Desires and expectations are projected into the distant years and are thereby freed from the contingencies of the present. Protected from the realities of a limited opportunity structure and accepting the myths of social mobility and personal "salvation" through schooling that are perpetuated in educational practice, school attenders continue to nourish unjustifiable hopes and unrealistic expectations. What will happen to their conceptions of the future when they find themselves, as the vast majority in our sample surely will, unable to fulfill their educational plans?

Among the adolescents we interviewed were 42 boys and 37 girls who had left school and were now looking for work or preparing for marriage in the more traditional sectors of Tunisian society. When in turn we asked these adolescents what job they expected to have at the age of twenty-five, their answers reflected a far more sober view of their future. While only 10 per cent of the boys who were still in school thought that they would have blue-collar jobs, 76 per cent of those who had left the school system now saw themselves in such occupations. Now, too, the distinction between hopes and expectations for the future is more clearly drawn: The "ideal" occupations the school leavers cite are significantly higher in levels of prestige than those jobs they expect to hold. Their empathic abilities seem nonetheless limited by what is realistically conceivable, for their "ideal" occupational goals are much less exalted than those that the school attenders envision.

Tunisia is typical of most developing societies today in the small number of suitable occupational outlets the economy can provide for its newly educated youth. The vacancies created by the departure of Europeans have long since been filled, and economic growth has lagged far behind the expansion of formal schooling. As a result, educational criteria for employment have become far stricter than the actual job descriptions would require. "Having completed primary school or having gone to high school for a few years," Allman (1972) wrote, "are no longer major accomplishments and sure guarantees of a job as they were only a short time ago in most developing countries like Tunisia." Indeed, the 1966 census revealed that almost 40 per cent of all Tunisian males between the ages of fifteen and nineteen were unemployed.

The "tragic situation" that Gustav Jahoda (1968) observed in Ghana is repeated here, namely, "that the widening of educational opportunities and improvements in educational practice increase the number of those school leavers whose hopes for reasonable jobs are doomed to frustration." Among the boys in our sample who were no longer in school, occupational aspirations were influenced neither by school attainment nor by socio-economic status. Instead, the significant factor was the number of years that had passed since they had left the school system: The more years these young men had spent in the frustrating search for a suitable occupation, the lower were their expectations regarding their occupational future.

In comparison with the school attenders, the boys who have left school have not only reduced their occupational aspirations, rejecting the dreams they too must once have believed possible, but they also have foreshortened their outlook on the future as a whole. When asked to list all the events they could imagine that might happen in their lives, 48 per cent of the boys in school looked beyond their middle twenties in listing their anticipations; but in spite of being a good deal older, only 19 per cent of the school leavers were willing to look even that far ahead.

In a country with high unemployment, jobs for women are particularly hard to find. The girls who leave school, as we have seen, return to their homes and their father's authority. There they are confronted with the traditional expectations tied to the role of a marriageable young woman. Secluded within their families, their schooling now seems largely wasted, their new perspectives and aspirations relevant perhaps for their children, but much less so for them. When asked to list the events that they thought would happen to them in the future, 50 per cent of the girls who had left school now spoke of marriage or children, and the vast majority no longer mentioned educational attainments or occupational success.

In comparison with the school attenders, moreover, these young women were able to name significantly fewer personal events that they anticipated in the future. Their occupational aspirations were not influenced by the amount of schooling they had attained nor by the number of years that had passed since they had left the school system. The decisive factor now shaping their conceptions of the future was their socio-economic status: The more educated and less impoverished their fathers were, the greater were their own expectations of being able to postpone marriage and acquire further training in preparation for skilled blue-collar or clerical positions.

CONCLUDING NOTE

From what we have seen in these last two chapters, it would appear that the time perspectives generally characteristic of persons living in isolated, "post-figurative" societies are rather readily transformed by the processes of modernization. The cyclical view of an unchanging temporal order, and with it the sense of a largely predictable future, post-figured in the patterns of one's elders, now seems to be a vanishing mode of temporal experience. As these societies become integrated into nation-states, and as the promise of modernization captures the imagination of people who once had a secure sense of timeless identity, the psychological "obstacles" that were thought to be major impediments to economic development—an "inability" to conceive of a different future, or an "obstinate fatalism" in the face of objective opportunities—generally turn out to be evanescent and relatively unimportant.

The sources of individual differences in outlooks on the future in the developing world, no less than in highly modernized societies, are now likely to be found primarily in the external circumstances of social and economic existence, rather than in presumed internal psychological shortcomings or in a culturally conditioned inability to imagine or strive for a better future. As Ivan Illich (1971, pp. 9–10) suggested, "Few countries today remain victims of classical poverty, which was stable and less disabling. Most . . . have reached the 'take-off' point toward economic development and competitive consumption, and thereby toward modernized poverty: their citizens have learned to think rich and live poor."

The Tunisian adolescents who shared their dreams and expectations with us seem consistent in the dynamics of their orientations toward the future with what the "preliminary perspectives" we developed in Chapter 1 would lead us to expect. The subjective future is linked by the end of adolescence with conceptions of past and present experiences, and it largely reflects the implications of present realities for future prospects. A person projects an image of himself into the situations that he envisions, experiencing in the present some of what he would experience were those situations actually to occur. If the emotions generated by such anticipations are unpleasant and people see little reason to expect more favorable experiences, they are likely to stop thinking about the distant future, and restrict their attention to the possibilities for gratification in the present and short-run only.

The Tunisian findings underscore the importance for occupational aspirations of remaining within the protective and myth-sustaining environment of the school. The young men and women who were no longer in school at the time of the interviews not only had far less optimistic outlooks on the future than the school attenders; they also revealed a deeper sense of the contrast between their hopes and expectations, a foreshortened perspective on the future in general, and a far more tentative belief in man's ability to control his destiny. These school leavers are in no way different from their peers who are still in school in terms of their cultural heritage or their ability to imagine the future. If they differ from the school attenders in the content of the anticipatory images they have created, it is largely because of the contrasting realities of their present circumstances and the self-conceptions and social roles that these circumstances imply. The major questions for Tunisia's future appear to lie in the gap between the psychological changes brought about by governmental policies on the one hand and economic and social development on the other, for the number of adolescents who leave school with new self-conceptions and personal aspirations far outstrips the available means for their smooth absorption into Tunisian society.

Characteristic time perspectives are indeed inherent in the social roles and class positions that mediate a person's relationship with the wider social forces shaping the direction of change in his society. In the next chapter, Theodore and Eleanor Graziano express in the crisis of their personal lives the converging impact of sex-role identity, work experience, social class, and social change—aspects of culture and social structure whose general manifestations we have now examined.

8
Theodore and
Eleanor Graziano

"TWO SEPARATE CLOCKS, EACH TICKING ITS
OWN SWEET TIME, EACH HEADING OFF IN A
DIRECTION THAT WOULD PROBABLY CONFUSE
THE ALMIGHTY."

Life is perceived as unfolding toward the future, and everything seems to culminate in a momentum, a thrust toward that future. Men and women stand behind the momentum, pulling the past and present along with them through their preparations, hopes, and imaginings, as well as through their recollections and regrets. Some of us attempt to resolve our conflicts and uneasiness about the nature of time through reconciliation or simple acceptance. This is the way it is, we say, and the way it must be. Others of us seek change, a new direction in our lives. We seek a break with the past, a break in time. Still others of us find change intolerable, for it upsets not only our prospects, but also the meanings derived from the past and present to which we are committed.

Indeed, cultural and social conditions make change in the flow and momentum of life difficult for all of us. Social structure—the organization, composition, and power of political, economic, religious,

218

and educational institutions—locks us into life styles that make it difficult to contemplate alternative possibilities. And despite the value we place on personal autonomy and individual achievement, the freedom to take effective control of one's life is, for the most part, a luxury afforded only a few in our society.

This last chapter explores further the effects of society and culture on people's perceptions of time. More specifically, we shall examine through a life study such issues as the role of emotion in shaping images of the future, the concept of possibility, the distinction made by Leach (1961) between repetitive action and progressive action, and finally, how cultural themes of time are incorporated into the social system of family life.

The conversations with Theodore and Eleanor Graziano that constitute this last life study reveal the ways in which culturally derived perceptions of time come to be lodged within a family. As we have seen throughout this book, the influence of culture and society on the definitions of sex roles, and in turn, the influence of sex roles on perceptions of time should dissuade us from conceiving of fatalism as primarily a psychological predisposition. Fatalism is also consolation, for there are those whose daily work at best accomplishes little beyond an avoidance of unpleasant experiences in the future. The realities they confront, moreover, require them not only to fight against hopelessness, but also to lower their natural ambitions. The future-oriented timetables of the upper-middle class that Roth illuminated are balanced in the lower classes by the relative absence of present and future stability, by the feeling Kahl expressed of having no place to go, and by Lerner's recognition of the revolution of rising frustrations.

How easy it would be to describe working-class people as present oriented. In truth, their perceptions of time are no less complex than those of any other group of people in a society. For despite the broad class differences in values, and despite the general truth of Niehoff's observation that "people will not willingly give up their old ways until they are well convinced that the new ones really are improvements," men and women constantly seek to change the logic of time and the context of inevitability that their age, sex, history, and culture have prepared for them. Granted, there will be some who seek a security in the past and rely on a future largely prefigured by their elders. For them, images of the future may well remain as "passive ruminations offering little sense of alternative possibilities . . . and little reason to intervene in an effort to shape a more favorable destiny." There will be others, however, as Ted Graziano reveals below, who take a different stance toward the future. Reconstructing the memory of their experiences so that the present and future might assume a new significance, such persons strive to transcend their own histories in order to derive the feelings that only a new future and, with it, new circumstances might yield. Mr. and Mrs. Graziano seek to resolve the discrepancies between their prior and present histories; between their own particular life styles and circumstances, and between their ways of managing the future.

I had been late arriving at the Grazianos' house on Poplar Street. Some unexpected traffic, which I had cursed, held me up, and that fact, coupled with the uncongenial design of Boston's streets, was enough to make me almost forty-five minutes behind schedule. Theodore Graziano, his good friends call him Mushy, had Tuesdays off from work and so the moments I could spend with him and his wife Eleanor together were precious. Normally, I would see one of them alone.

For about fifteen years, Ted Graziano has worked for a Boston newspaper. Starting as a stockroom boy (he was almost twenty-five at the time), he had worked his way up to where now, at thirty-nine, he was foreman of the shipping operation. Salaries had risen over the last years. Several threatened strikes and union pressure on management had brought his take-home pay to a level at which he could just about get by. The important thing for him was the security, a factor he had mentioned several times when we spoke in his office:

"This is not the time to be sitting on any gamble," he said as we began eating our sandwiches. "This is the time when the economy of the country demands that you get yourself a job that looks like it's got to hold out. Doctors, lawyers, judges, they got it best. The working man, like always, he's going to be the first to get hit. Unemployment starts at the bottom and works its way up. It's like a disease that they can't find a cure for. You know it's out there and you just got to do the best you can to avoid getting it. Stay warm, eat good food . . ." He smiled at me, enjoying his humor. His strong hands held the sandwich he had made that morning. His black hair was thick, his forehead low, and only the thin lines around his eyes and mouth revealed his true age. It was hard to believe that Ted was approaching forty.

"Yeah, forty," Ted reflected when I raised the issue with him once. "I guess that means something. I still feel young though, and you know what they say about feeling your age."

"Yeah, I do. But I feel old."

"Maybe you ought to see a doctor," he laughed.

"Or stay away from young-feeling guys like you." I remember that particular remark. It had sounded patronizing, but he had enjoyed it. Still, the more I played out in my mind a conversation that might have ensued, the more patronizing it became. I think often, for example, that maybe the "good life" is working with one's hands, witnessing real accomplishment, and going home with no lingering thoughts about work until the next morning when you get up and repeat it. But that is precisely the point on which Ted Graziano had long ago put an end to my naïve notions.

"You take a job like this," he had said, waving his right arm about as if to take in the enormous shipping area, partitioned offices and truck docks, "it's one helluvan operation. I got real responsibility here. You ask any man they got working here and they'll tell you about that responsibility. I tell you though, in the beginning, ten years ago or so, it was one helluva challenge. Couple nights there I got so damn excited with the prospect of it all I could barely sleep. Now, it's just another job to me. Plasterer gets up each morning and slops that shit on the walls; butcher goes cutting up his meat; I direct the newspaper business down here. It's a job, not much more. Every day you get a little excitement, but when you go to think about it day in and day out, it's routine. Man, there are times I'm working down here thinking about how nice it would be to be anywhere else. Sitting in the sun or taking care of people. Doing what you're doing, you know, walking around, talking to people. That'd be nice. I'd like that. But most of the time I have to take those thoughts and crush 'em up like little paper balls and throw 'em away." He shook his head from side to side, crumpling an imaginary sheet of paper in one of his hands. Then he looked back at me.

"Tell you what gets me—I know you're interested in this. It's the time thing. Two things about time I can tell you. First, is how slow it goes most of the time. Look there, once you start to use that word it just keeps coming back on you. Lots of ways you can use that word, though. Like wasting time. Boy, I must say that to the guys here twenty times a day." He had heard the word again and grinned. "Well, I ain't about to go commenting on my verbal skills. Wasting time is one thing, but it's not as important to me, really, as the way time passes so slowly most of the time. Clock seems to die on me every day. I'll be working and just check it out, then work long and hard and check it out again, and see only about fifteen minutes have passed when a good hour should have." He smiled. "You can't beat it. You just can't beat it." A feeling of resignation could be heard in his voice. The muscles of his powerful forearms moved slightly as he pulled his watch band higher on his wrist. "Here's the devil," he said. "Right here's our number one nemesis. Nothing I can do, it seems, to make the seconds go by any faster. 'Course, I'm sure that in a few years I'll be wishing I could slow the whole mess down."

"Yeah," I responded, "I guess lots of us know that feeling right now."

"Funny, I don't. Not yet anyway. Probably will pretty soon. Ain't getting younger just sitting here, but, like, you could say I'm still in the stage of life when I want time to speed up. You know, you could probably categorize people just by asking them, 'Do you want time to speed up or slow down?' I'll bet you'd discover some pretty interesting things. Of course, you'd have to take account of the fact that lots of times a person is doing things where he wants time to speed up or slow down. Still, I'll bet you'd find that lots of people feel the same way most of the time. Like me, I want it to speed up. It's true even when I'm home."

On another occasion, as we sat together on the concrete edge of the truck dock, with the last truck having departed and Ted ready to head for home, he offered another thought about his involvement with time. As usual, his tone with me was formal, for he never forgot that I was not one of his close friends.

"Maybe the worst part of the job is not that time passes slowly. The worst part of my life, really, is that I can see the whole thing laying out before me. See, if I were to tell you that the job is boring, monotonous, that wouldn't really capture the feelings I have when I think about it. And man, you *make* me think about it." He slapped my leg. "I always feel talking with you that I'm back in school writing one of those themes for an English teacher, and I'm trying my best to dig out the big words. Funny about that, too, that being back in school, 'cause like right now, trying to answer your questions, I want to give a good impression. . . ."

"Like getting a good grade?" I interrupted.

"Just like it. But there's more, too, because when I wrote those themes, or tried to write them," he smiled, "it was like I was talking to myself in a very deep and special way. This is all, now, different from working for a good grade. When you write, you know, you begin to know yourself in private. It's not like talking." I could see his thoughts beginning to overtake him. Perhaps he was saying that tenderness as well as anger can come out in writing. Hurt can, too, in ways that day-by-day encounters and friendships often do not allow.

"Anyway, let me get to this one thing. The rotten part of my job here is that everything is so predictable. I can see long, long stretches of time. Maybe the next ten, fifteen, twenty years. I can see them. Work and work, the same job. That's the monotonous part. If every day were the same, like it is, it would still be all right if I didn't ever have to wonder about how all the days string together. But it's the line of days, one after the other, each one repeating, and then the ability to be able to look down the road and see exactly what's coming. Jesus, that's, that's. . . Honest to God, man, it just about frightens me to death, because it means I can see the days leading right down to the end.

"You know you can't live every day of your life thinking only about that single day. That's part of the crap they feed you. You can't do it. You can't keep your mind from focusing on tomorrow and next week, next year. That's what I'm thinking about. You don't think that way in this life. You've got to think about the future. That's where the possibilities are. That's where anything that might be resembling hope is. This business of living day by day, even in a job like mine, can't make it. You can't make it that way. So now I got the problem of being born with a vision that looks down the road, way down the road, and being able to see everything that's coming. They got lots of guys, I'm sure, give their right eye to be able to see what's coming up for them. Well, I can see, and just being able to see is more of a curse than it is anything else. An evil curse. You know what I mean? You got this inborn thing that drives you to get yourself set for the future, right?"

I nodded yes. I thought, too, of the connection between one's images of the future and one's present mood. For when things go particularly well I am able to foresee the future—almost as if the sense of possibility cannot be contained by anything as finite as a future allotment of time.

"Yessir. Every man does," Ted went on. "But you don't know what's going to be. Today you talk with me and I tell you something. Tomorrow maybe you talk with Ellie or some kid and you don't know what they're up to. And in a week from now we may not even be seeing one another. You'll go get yourself another family or go do a story about something else. But you can be sure as hell I'll be right here. Two weeks vacation in July each year and then I'll be here. That gets to a guy, that knowing that you might be moving up, like I have, in an organization, and still you're not actually changing that much. I think that's why so many of us liked being in the army so much. Didn't anybody want to get killed naturally, but it *was* a change. Everything that led up to the army stopped once I got in, and what would come after no one could see. I thought about my future plenty then. Oh brother, we had a million conversations about the future. But no one could tell us the way it was going to be. The future was all a mystery. I remember, that was my word for it: 'the mysterious future.' It made you kind of scared. But now that I think about it, those little jittery feelings were exactly the feelings I needed to get me going. They give you a kind of a push, a motivation. Now, my whole life is like a car running along a highway. Things going pretty smooth, no complaints, but pretty soon the scenery gets monotonous. So what do you think about? You think about the way it used to be, or maybe you start to wonder just how long it's going to be before your car runs out of gas." He stopped talking, although the movement of his face told me of the thoughts that preoccupied him.

My own thoughts, however, were turning to this method of inquiring about another person. Despite the care one takes, it is clear that many questions or concerns simply do not touch the lives of those with whom one speaks. Nonetheless, in response, they work hard at providing answers. And so those whose lives we may or may not gently invade transform disinterest or irrelevance into unpreparedness, and communicate to us that it is their fault and not ours that a gracious or beguiling response is not forthcoming.

For Ted, the subject of time was central and preoccupying. I had not fished for inquiries that might, somehow, conjoin his world with my own. I had, as they say, struck a chord, a reservoir he had not only wondered about, but seriously considered. I would hear his reflections on time again on the afternoon I reached the Graziano's forty-five minutes late. His orientation to the future, and what he meant by a transparent future, were not capricious notions for him.

In particular, a temporal orientation predicated on personal experience was forming even in the midst of the random flow of those conversations that often seemed to me to be heading in no discernible direction.

Our home meetings took place in the kitchen, so I entered through the back door as the family members did. Alone in the house, Ted had been sitting in the kitchen reading the newspaper. Two glasses of milk and a cake covered with tin foil were on the kitchen table along with plates and forks. Early in our friendship, Eleanor Graziano had learned from Ted of my love for apple cakes and apple pies, and so something freshly baked regularly awaited my arrival. Today the smell in the kitchen was especially tantalizing. Ted watched me and grinned with pride. "Quite a lady, huh?"

"You better believe it," I replied.

"She'll be here in a minute. Something came up at her mother's. She left a note." Ted handed me a page from a calendar on which Ellie had written: "I will be back in a minute. Milk and cake are for you. I have not forgotten."

"Come on, let's eat the cake. I've been waiting for you dying with hunger here." Ted took great pleasure in cutting the cake and gently laying a piece on a plate. He reached over and placed a fork on the plate and then pushed one of the glasses of milk toward me. When I was taken care of, he prepared a piece, considerably smaller, for himself.

"So what do we talk about today, or shouldn't we say anything until Ellie gets here?" Ted asked, his mouth full of cake.

"Talk about anything you like, Mr. Graziano," I said. "I'm not listening anyway with this cake here."

"Eat it up. There's lots more where it came from. Your wife cooks, though, doesn't she?"

"Yes, she does."

"Well, then you aren't hurting too much." He sighed deeply, wiping some crumbs from his chin. "Lots of guys have it pretty bad. I told myself a long time ago, though, no wife of mine is going to work. Ever! No matter how bad it is, a man provides in his way, a woman in hers. Ellie doesn't need to go cleaning or secretarying, or work at the phone company like her girl friends. We'll manage. Eight thousand years, and I'll have this house paid off, and when I die she'll be set up. She don't ever have to work if she don't want to. That's the way it's supposed to be. You agree?"

"Well, I guess I believe people have to work it out the way they want," I answered him. "Men live their way, women theirs."

"Now you're not going to tell me you're one of these liberation people, are you?" He smiled at me.

"Well, I guess in a way I am. Men and women, it seems to me, have to enhance one another, support each other to be what they want."

"Now, *that* I like. *That* I'll buy. That's all right. I like that enhancing business. You know, now that you mention it, that's maybe like my biggest gripe. I never did have anyone enhance my life. I can look back at it now, forty years next July, I can't see a person there anywhere enhanced me, helped me to get anything. Parents never gave a damn. When they died it made no difference in my life at all. No teachers ever cared about what I was doing in school. Priest couldn't have cared less about the whole group of us kids."

"You have Ellie," I said.

His eyes were closed slightly as several thoughts seem to touch him at once. "Yeah, I got Ellie, all right. Probably a good thing for me, too. But it's not really enhancement. That's not what I call the kind of support I'm talking about. A wife is a wife, she doesn't help you out there." He pointed in the direction of the kitchen window above the sink. "That's not what I mean."

"What is it then, Ted? Can you talk about it?"

"Let me get the words first." He paused, looking through his glass at the filmy layer of milk on the bottom. "The living day by day that women do is not what helps a man, or at least a man like me. There's a way a man has to get his whole life together. It's all got to fit somehow, make sense. You know what I mean?" He didn't wait for an answer. "It's got to be set up so that every move you make has some reason in a plan that you have to formulate somewhere along the line. It has to build toward something. Even when you're a little kid, people are asking you what you're going to be when you get older. I did the same thing with my kids. I remember holding Tony on my knee one night, I'll bet he wasn't six months old, and telling him about some of the things he might think about doing when he gets older. Now, here he is, what? Fifteen? He still doesn't know what he's going to do or be. But it doesn't make any difference, see, 'cause he knows he has to become something. That *I* put into his head, starting when he was six months. That's the important thing. He knows he's got to have a plan for when he gets older.

"My problem maybe was I never had one, really, although I did have this thing about making sure that you're living your life to prepare. That's about the only good thing my Dad ever did for me. He made me understand that you got to have something out there waiting for you, and that nothing is going to be out there unless you make the preparations, do the ground work, like they say."

"It makes sense to me."

"Sure. It's got to make sense to you 'cause you're a man. You know about these things. You may be richer than I am, I'll bet you are, too," he looked at me, "and you may have a more interesting job, which I know you do. I even told you once, remember?"

"Yes. In the shop."

"At the office, right," he corrected me.

"Office," I said quietly.

"But the important thing is that a man knows what it means to arrange for his life. That, no woman knows! That I know for sure." A melancholy feeling was taking the place of anger. "There's a kicker in it though, you know?"

"Which is?"

"Which is that a life that asks us to make sure that pasts and presents and futures fit together in some logical way has a price. You know, yourself, as time goes on it gets harder and harder to just live each day and get the most out of it. Can't be done. You don't live each day like that when you're a man. What you live is your work at the moment, your plans, what you call your prospects, *and* your regrets, what you should have done. You think about having bits of time back again to work with. You think how nice it would be to see what the future holds, even in a job like mine where I practically know how everything's going to turn out. Still, you'd like to take a little peek. Maybe just to know whether all those connections you're making are really sticking. You follow me?"

"Yes, I do," I answered him.

"So you don't say the hell with what was and what will be. Like that song, what will be, will be. That's the way you think when you go out drinking. At least, that's the way it is for me. I go out and get a little in the bag," he started to smile. I smiled too, as I saw his face take on a look of pride. "That's when my future disappears and my past disappears and I got once and for all my present moment and nothing else, goddamn it, to be concerned about!" He shouted the words as he straightened up in the chair. "But now, I ask you, how long's a man with any self-respect going to run around in the bag? So you come home, and you sleep, and you drink coffee, and you make nice to your wife, and there you are the next day, thinking the same things all over again: how it all fits together, how you wish you could be young again, and how great it would be to know what the future's got waiting for you. And there you are, right back where you started, if you got any strength, that is."

"That's one helluvan analysis," I said, moving about in my chair and wishing, for the moment, to take a second piece of cake.

"Well, I don't know I'd call it an analysis, exactly," he said. "Nothing mysterious about what I've been saying. You keep your eye on the future. That's what men are for. You keep an eye on the future because the name of the game is that since you don't have much to say about dying, you have to fill in as many of the empty spaces out there as you can. You don't leave things to chance. Ellie always talks to me when I get in a huff about God or fate. 'Things will just happen,' she says all the time. 'Things are just going to happen.' Sure. They are. Lots of things are just going to happen. Like the roof might fall in on our heads, right here in the kitchen. But the odds aren't pure chance. The odds are against it because *I* fix the roof, *I* make certain things like that *won't* happen. That's what you do with the day. You fix things so that some things will happen and some things won't.

"The difference between Ellie and me is she waits. She waits for the roof to fall in. Then, when it happens, she does something. Or, she'll say, there's nothing I can do, so I shouldn't worry. What are you going to do? Men worry. They worry because they have to reorganize their lives to make the future better or safer or whatever they're after. Women bake cakes. You like her cake, she's pleased, that's enough for one day. Men worry too, because they have a plan which they have to put into operation." Ted rose from the table and went to the sink to wash his hands. I remembered being in this house one afternoon when the sink faucets were leaking and his temper, like my own on many occasions, had exploded. "No, now they see the operation taking place. Maybe it works, maybe it doesn't. The point is that with one part of their brain they're wondering what they could do *now* to make next week or next year work out better. They're starting new plans, if you know what I mean." He dried his hands on a blue towel that hung on the handle of the refrigerator door.

"Now with the other part of their brain," he resumed, seating himself opposite me, "they wonder what it would be like having the past back again so they might fix a few things, maybe play it all out a little differently. But what a man does that a woman doesn't have the faintest idea about is, that when he has the chance, I mean, *if* he had the chance to live it again, he's still working it around so that the present might be better. I should have gone this way rather than that way, he'll say. Do this instead of that."

"I think I do that myself."

"Sure you do. But you know what a woman does? She gets a chance to be young again, she'll just do it all over the same way. You ask Ellie. She'll tell you exactly what I'm telling you now. All the things from the past that she liked, she'll just do them all over again. And all the bad things too." He anticipated my question. "She'll just say you have to live through the good things and the bad things, or whatever that expression is. Take the good with the bad.

"You know what it's like, Tom? It's like football. There's a good reason why men like football. It's not just the violence. It's a man's game because it's played the way men think. Look what you got in football. First, you got competition. You got a chance to win or lose, and a man is practically born with the idea in his head that you got to win. Then, everybody's got a fair chance. But now here's the subtle part of that game. You have a game plan. Plan. See, there it is again. Now, this plan takes in your purpose, your goal, the way you're going to work it out. And you stick with it until you have to find a new one. And that's the point right there. You don't scrap a plan. A new plan simply means you have to find new patterns, new ways for things to stick together. And it's all leading up to something. Something you could practically predict. That's what football is. If you move here, I'll move there. You move here, I'll move there." He pushed his fingers against the table top, his right hand representing my team, his left hand representing his team. "You can make predictions. It's not a game of chance, like gambling. You know how something ought to work out. Here's another thing. What do they do when the game's over?"

"I don't know. They think about what's gone on."

"That's exactly right. They see the movies of the game, the replays like they call them. They use those films to change their plans and, in your words, to try to make the future come to pass in a different way." He paused. "Here, I got some more milk."

"That's a great metaphor, Ted. I've never thought of football in those terms. It's absolutely true."

"I guess I've been thinking about ways to carry our last discussion from the office forward a little bit. I can't say these things to Ellie. Of course, the kids don't want to hear about it. I talk with some of my friends, but, as you can imagine, not too many of them are as smart as you. I guess you know better than most of us how little schooling we all had. Our kids, I hope, will be different. Oops, I'm starting to sound like the old lady. Hoping instead of arranging."

"Well," I tried to assure him, "there's got to be a place for hope in the plan."

"Sure," he said with resignation, "of course there's a place for hope. But first you make preparations, then, whatever is left over that you can't control for, that falls over into the hope category. A fellow in my position ought to be able to say without fail, I can predict that my kids will go to college. I will get money somehow for them to go, whether it is good for them or not—and, believe me, I got lots of questions for you on that one—I can arrange the future for them. I don't need God or prayer, or hope. This is exactly what I'm talking about. I can arrange for all of this."

He rested his elbows on the table and held his hands up as though shaking a loaf of bread. "I can get all of this to happen. Fate's got no place in here. But baby, I'll tell you, sometimes it makes me wonder why we're here, and why we're doing what we're doing. You got to get out once in a while. I don't mean, don't misunderstand me, I don't mean just out of the house. . . ."

"I know what you mean."

"I mean out of the path of the ball, you know? Go this way for once, or that way, any way to break the monotony. Go backward or forward if that makes any difference. That's what I mean by getting out. In time!"

For a moment, Ted was silent. Then he resumed. "Prediction. That may be the key. Each man his own computer, able to see what's coming up at any point in his life. There are lots of guys who tell you that you can't go around predicting. They say it's bad luck. They're superstitious. Predict something good about your life and you'll see, something bad is always supposed to happen. That's what they say, anyway. Ellie goes along with this. Only God predicts, is probably what she'd tell you if you ever asked her about it. Now, don't get me wrong. You don't want to go around predicting. But predicting is like I was telling you before. It's like hope. It's part of what's left over. You have plans that you might like to see how they come out, but I'm not so sure you have to go around predicting. Still, like that football thing of ours, what'd you call it?"

"Game plan, you mean?"

"No, that other, like the picture I was describing."

"Metaphor?"

"Metaphor. Football metaphor." He seemed relieved. "I remember that from school. Similes and metaphors. Right?"

"Right."

"I never got straight which is which. Anyway, prediction. When you get to be a man they've got this future thing so pushed down your throat, the feeling of making a prediction just isn't so weird as it seems at first. Anything like a . . . technique that has to do with working out the future is something a man quickly gets used to. Not women. Hope's still the last resort though. Just like they say. When all else fails, maybe you better start praying. But when they say that all else fails, they mean that they've been working at something that isn't going to work now. You don't start with prayer. You may see those football players pray at the start of the game, like to get a little boost from the Good Lord. That's all right. Little superstition never killed anyone. God, Ellie better not hear me say that. They may be praying, but you and I know they've been practicing all week long. Not praying! Practicing! They've been getting themselves all set for Sunday. So if they make predictions, it's because they feel they have ways of fixing the outcomes. You get what I mean?

We had been speaking almost an hour when the back door opened, and Eleanor Graziano entered the kitchen. Her face was flushed and she was out of breath.

"Please, please excuse me. I'm so late. I went over to my mother's for a minute. I didn't even take my coat off and before you know it an hour's gone." I nodded to her and started to say something about my own lateness. Bracing herself in the doorway, she leaned down to remove her boots. "Did you find the cake?"

"Did we!" I replied.

"Was it all right?" she asked modestly.

I grinned at her. "Fair. It was fair, Ellie." She laughed and turned her face downward. Ted laughed too.

"A couple of jokesters I got here. Two little boys with no place to go and nothing to do but eat cake and get fat."

I couldn't resist: "Oh, you love it and you know it, Mrs. G."

"You I love," she laughed. "Mushy I'm not so sure about." Ted and I laughed again and looked at each other.

"You better be careful, wife, or I'll tell you what I've been telling Tom here for the last hour about *you*."

"Yeah, and what's that?" she asked her husband, throwing her boots in the back hallway.

"Oh," he started, "about people enhancing one another's lives. Game plans, you know, the usual." He winked at me.

"Game plans?" she asked. "Football? Is that it?"

"That's it," I said.

"Not quite," Ted said.

"Then what is it?" she wanted to know.

"Sit down, Ellie. You and Tom talk. I'm going to the corner for cigars. When I come back we'll continue with this. It's time for a commercial."

"Commercial? Tom, will you please tell me what he's babbling about?"

"I will."

"Go on then," she said to her husband. "I've got a few things for Tom *you* don't have to know about."

"Yeah, like what?" Ted demanded.

Ellie was silent for a moment. Then she exploded the one word. "Hockey!"

The three of us roared with laughter.

Eleanor Graziano moved about the kitchen, looking into cabinets at canned goods, glasses, and plates.

"Will you stay for dinner?" she asked.

"No, I can't, Ellie, but I promise I will another time."

"Can't seem to find . . . don't tell me I forgot to get . . . if I don't make a list when I go to the store I'm dead. Can you beat this? I'm not forty yet and my mother, who is almost seventy, has a brain in better shape than mine."

"I doubt that."

"No, it's true," she protested without looking back. "So what do we speak about today? Are you still asking the same questions?"

"Same questions, I suppose."

"About working and living and everything, eh?"

"Yeah, that about covers it."

"Well, I don't know what else I can tell you. Mushy's the one with all the answers. There's nothing you could ask him that he wouldn't have an answer for. Everything. Everything he can find something to speak about. Some man, Mr. Graziano."

I watched her lay out supplies and utensils in preparation for making dinner. The counter tops were spotless, the sink empty of dishes and garbage, the faces of the cabinets glistening clean. It was all quite a change from the day when the Grazianos had moved into this house. Four years ago they lived in a four-room apartment less than a mile away. Then, suddenly, Ted had gotten it in his head to buy a house. Ellie argued that they couldn't afford it, but Ted was driven to buy, and with loans and arrangements at work, they managed to purchase this lovely three-bedroom home. Now they "could spread out," as Ellie said, finally admitting her delight. "The children can stay in the same school and we will live better. I'm closer to my mother, too. If we can just manage," she had sighed. "If we can just manage it will be a blessing, *the* blessing of my whole life."

Although Eleanor Chadwick had never known poverty as a child, she also had never imagined that her marriage to Teddy Graziano, "the Mushy man," would ever lead to a home and a kitchen glowing with warmth and pride like this one. Her father had been an elevator operator and starter during the day, a warehouse inspector four nights a week. He had lived his entire life in Boston. His salary allowed his wife and five children to live comfortably enough in a three-story walk-up apartment. A screened-in back porch opened out from the kitchen of that apartment, and on hot summer evenings the family gathered there feeling as cool as one could during a Boston summer. The view from the porch admittedly was rather uninteresting. "Who, after all, likes looking down an alley at a bunch of other apartment houses?" Ellie had said. "But it *was* cool up there. We enjoyed it together very much. That was ever so lovely, just to be together, even when it was so hot you thought you might die. They had a kid, I remember, who lived downstairs, who died of the heat. He just couldn't get enough cool air, the doctor said. So tragic. So tragic. He could have come up to our porch and saved his life. I suppose you could say I long to see my childhood again."

We had had this conversation about her childhood several years ago, on the very day, actually, that the Grazianos moved into their new home. Ellie and I had sat in the kitchen on cardboard boxes filled with supplies and clothes. "I would love to have those days back again," she said on that moving day. "Even before Mushy and all this. So many times I wish that tomorrow would be the beginning of a change, a sort of a change backward. It's not like you might think, that I fear getting older. I don't think I'm different from other people. But no one likes to grow old. When you have children they say it keeps you young. It's not true. Children age you. My, they age you. It's just that they don't give you any time to think about it. They got you running around all the time.

"No, it's just that wonderful feeling of worryless peace, childhood, being a little girl and having those long hours with my father. He was a delight to know. You would have liked him, Tom. I feel that very strongly. You two would have gotten on. 'Course, that's not saying too much, since Dad never had an enemy. He got on with everyone he ever knew. You know, in many ways, he was very much like Mushy; Mushy without the anger. Dad knew, for example, that he would never be rich, that he'd never achieve anything special or wonderful in his life. He knew lots of rich people, too, but I never heard him complain, or compare himself with someone else. The world was the way it was, and he was big enough to accept it. All of that, I'm sure, made for very peaceful living for the rest of us.

"Mushy's the same way, up to a point. I'm sure he feels that he's not going anywhere in his life, at his job, I mean, but he's not about to accept any of that. He wants to achieve something, to be able to show everybody, mostly himself, of course, that he's gotten something out of life. He'll always tell me that if next year's the same as this year, then he's been a failure. There's no other way. That kind of statement I never heard once from my father's mouth. Never once. For me, you see, that's the sign of a good man. If next year's the same as this one, then you thank God. You get down on your knees at night and you thank God that everybody is well, or still alive I should say, and that you have enough to eat, and a comfortable place to sleep. You know that everything is provided for. But Mushy doesn't want it that way. The comfortable things, of course, he wants. That he wants as much as anyone. Like, he wanted this house. . . ."

I remember Ellie looking about at the boxes and cartons and seeing in her face a look of "How are we ever going to get set up again? Why did we ever undertake this move and this severing of attachments?" I saw excitement, too, a controlled excitement as though one were not supposed to ask for such delicious treasures, but could nevertheless enjoy them if they came one's way.

"Well, I'm set here for life. This is it for me. I detest moving. I may be strange or peculiar, maybe I'm some kind of a mental or something, but I become attached to places. I get used to something very quickly and I don't ever want to change. You know something? I can feel in my hands the curtains we used to have in the apartment where I lived as a child. And the tables in the living room and the kitchen, too. That's how strong my memory is, so that must be how strong my attachment to things is. To people, too. It's sort of funny, though. It grieves me, it really does, that the good times are gone and by many, sad to say, forgotten. I think Mush is that way. Something good happens, like maybe a party. He likes it as much as anyone.

But then, when it's over, that's sort of the end of it. Yesterday for him, even this morning, might just as well be a thousand years ago. Me, I'm so different from him it's hard to believe we could have gotten along as long as we have. Something important happens, it doesn't matter whether it's good or bad, I hang on to it. I don't ever want to let go. It's just like the house. I wanted to hang on to the old house. It has memories in it. That's what he doesn't understand. I love this, of course, but you give me the largest mansion you can find, a palace, and I'll sit in my bedroom, I mean bedrooms," she laughed, "and I'll be thinking of that little apartment I grew up in, and the porch and the summers, and all the rest of it."

She paused, looking at nothing in particular, screening a series of ideas before selecting one to present.

"You know that my mother will be seventy years old in March? That's even hard for me to believe. It's all so peculiar. I think peculiar's getting to be my favorite word. But it *is* hard for me to believe that she's that old. My mother is always forty. When I think of her she's just forty. She's in that old kitchen—God, that was some horrible place—working at making dinner, or cleaning out the stove, or peeling apples."

"And you're about ten?" I reasoned.

"Yes. I'm ten or eleven, watching my mother and admiring her very, very much. She was a beautiful woman, though no one could see that now. Beautiful, and blessed with endless energy, too. She's still got that, even now. But you know, what's funny is that when I think about her I'm not really remembering those years; I'm actually back there. I'm *really* ten, and she's *really* forty; that's the way it's always going to be. Neither one of us will ever change."

I could see in Ellie's face a sense of peace that bespoke the loveliness of those moments thirty years ago. "You see, I do become attached to places and to certain people and find myself hoping things never have to be any different than they are right now. I remember on many occasions as a child wishing that I didn't ever have to get any older. I even spoke to my mother about it. She always laughed at me. 'You're too young to be thinking about that,' is about all she could say to me then. Now I'm getting close to her age. Three years and I'll be forty myself, and I have a daughter and . . . well . . . life goes on."

That was a part of our conversation of almost four years ago. Now, as we waited together in the same kitchen for Ted to return, I know that we both felt a nostalgia, a longing for something, for the people of four years ago, for the people of our childhoods as well.

"You're pretty deep in thought, young man." Ellie surprised me. "Maybe *I* ought to be asking the questions today. Turn the tables on you." She was smiling.

"Yeah. I guess I was pretty deep into something there. I was remembering my own childhood, and also the discussion we had in here when you first moved. Remember that?"

"Yes." She strained to recover some of those earlier words. "Barely I do. About my childhood and the old place? Isn't that what we talked about once?"

"Yes. And your mother and father," I reminded her.

"Yes. I remember. My days as a little girl I certainly remember. I wonder what Mushy remembers of those days. I'll bet you very little. He sure likes to lock his past away. He thinks only of the future now. That's all. Every night in bed he's got another plan, another dream. Someday we're going to do this; someday we're going to do that. Trips, property, real estate, the kids going to college. This one's going to be a doctor; that one's going to be a nurse. I have to tell him, 'Mushy, it's 1970, not 1990. Let time pass already. You're living in an age that isn't even born yet.' Can you hear me saying that to him? 'Give the world a chance to do what it's going to do. What do you want the time to pass by so quickly for, anyway?'

"You know, I begin to think that people who don't have parents want the past to go away, or they avoid thinking about it, somehow, by doing all that planning and preparing. It's like today. I went by to bring my mother some groceries. Days like these it's hard for her to get around. So I brought over a few nothings, a little soup and some coffee. Bread. The usual." She blew out a long breath. "Anyway, when I saw my mother I couldn't help thinking how really old she looks. I see her regularly, almost every day because we live so close, so I can't see the changes taking place that easily. You need to stand back from it. But today, for some reason, maybe because I knew you were coming, I looked especially hard at her. And she's old! It frightens me. It actually begins to scare me. I'm not like my husband, with all his plans. I don't want it to be 1990. I don't want to be thinking about what's it all going to be like then, if we're still here. I don't want to go back, exactly, but I certainly don't want it to be 1990 or 1980 or whatever the king up there is thinking about at night. He wants it to be 1990, that's fine with me, but he can damn well get there without me. I'll just take my own sweet time about it." She ended by pulling extra hard on the tie of her apron strings.

"Ellie, what did you mean, you were thinking about your mother getting older because I was coming this afternoon?"

"I knew you'd ask me that."

"I knew you knew." We smiled at each other.

"Well, you get me thinking, somehow, about time, only I can't always find any complicated ideas in all of it. Time for me, like I told you once, is people. People get older, never younger. You reach a point in your life when you first begin to realize this and then you worry over it. Worry's maybe not the right word. But, you know, you feel bad that your little baby's grown up and is a big girl or boy, and that your mother really starts to look old. There's no denying it. I saw her today; it was almost like I was seeing her for the first time. She's an old lady now. She looks just like these old ladies you see wherever you go. When I was young, my brothers and sisters and me, we used to be playing somewhere and we would laugh at old men and women. I remember that like it was yesterday. When you're young you never think that you'll be one of those old people someday. You never do. You also don't think that much about your parents getting old. Older maybe, but not *that* old that you'd look at them and laugh. It's funny too, because I remember then, I still do it now, I thought about what it would be like when my parents died. Every child, I'm sure, thinks about things like that. But you don't think about them getting older or looking very old. Shriveled up and all, I mean. At least I didn't. Maybe I'm different. Oh, it's an awful thing to be talking about." She shook her head as if to be rid of these images.

"Well, anyway, just to finish, I make people my calendars, you might say. To see their faces is to see time. You can tell the date by looking at your mother's or your husband's face. You know what I mean? So today when my mother opened the door, I was flabbergasted. Your questions always make me think of the past, of my parents, and going to school, or whatever pops into my mind. I must have been thinking of my mother as a young woman. Forty or so. I think that's all I mean. You know, this is what you and I once talked about. My special way of not remembering, exactly, but of *being* the age of my memories. When they took place, I mean. That's what it was, I think. I'm not crazy, am I? I mean, this sort of thinking doesn't mean that I'm ready for the State Farm already, does it?"

"Well, it's either the State Farm or making apple cakes every day when I come."

"You want some more? Where did he put it, that miser? Funny, he's so busy planning he forgets to think that maybe you'd like some more cake."

"No thanks, I'm full."

"You're not," she said, "but I'm not one to force anybody."

For a while we sat together in silence. Often it feels awkward being in the house of someone with whom one shares intimate things, stories, memories and all. I feel that I should have questions ready at hand, for the other person may take my silence to mean that I'm not doing my job well. They're waiting for me, I tell myself, to keep the time filled and to take the lead. I started to rise from my chair. Through the window above the sink I saw Ted walking through a light rain. Ellie could not see him.

"Here he. . ." I started.

"I know one thing that makes me different from him." She had one last thought, one that Ted would hear later, but that I was meant to hear first. "I worry more than Mushy. I worry about what will happen. I worry there may not be ways to get things done as we want them to. But I don't worry about what I can't see, or about what's not here yet. What's more, and this is very important if you want to really understand us, is that I believe you must live with a belief in God. Some things you can't take care of yourself. Some things only God knows what to do with. Mushy thinks I'm nuts when I say this. So I've stopped saying it in front of him, and, for that matter, in front of the children when he's around. It drives him crazy. He worries, too, but for him it's that there won't be enough time to finish all the things he's planned to do. As for God, he's about as far from God as I am from, from, I don't know what. The President of the United States. It's not easy living with this disagreement either." I saw the top of Ted Graziano's head pass under the window.

"Don't think for one minute I'm telling you little petty businesses in our lives," Ellie continued. "I'm telling you the most important things. This is a very serious difference. We run, you might say, on different clocks. You know, like the East Coast and the West Coast. We're in the same country, speaking the same language most of the time, but we've got clocks inside of us that I'm sure not only run at different speeds, but, if it's possible, in different directions as well. That probably doesn't make any sense to you, but I think it sort of tells what I worry about quite a bit of the time." She hesitated again, although she had more to say. The back door flew open and Ted entered the kitchen.

"A man with cigars," he began, "is a man halfway to heaven." He slammed the door shut, and the latch chain smacked loudly against the glass pane. Shaking off the rain and stomping his feet dry, he was a man doing his best to announce his arrival.

I had turned my body to greet him. Beside me I heard Ellie say quietly: "Two separate clocks, each ticking its own sweet time, each heading off in a direction that would probably confuse the Almighty."

How and when the tension between Ted and Ellie grew to the

proportions it would that rainy afternoon I cannot recall. I do know that their ordering of the world, the arrangements they had made with time itself, were now at stake, driving them toward some temporary conclusion, some partial settlement. Just beyond the boundaries of this extended moment lay a special and distinct instant when confrontation would switch off, and the flow of this one eruption would, like water, find a new bed in which to run. No death, no final stopping place, but a life marker, nonetheless; an instant colored with hatred, I suppose, and the anguish that life and people, circumstances and history, combine to create.

In this case, I was at the center of their tension, for I had not only asked the questions on which they had unloaded so many feelings, I had split them apart by speaking intimately with each of them. I knew things about each, they both believed, that they had not confided to each other. Mine was the easy position of being able to listen to someone's everyday world without having to share much of my own. It is that comfortable role of sharing compassion while hiding one's own conflicting or antagonistic attitudes. Ellie was right. People are themselves temporal units, delicate mechanisms in which the experience of duration and the capacity for knowing change, the sense of history and continuity, disruption and evolution are all curiously combined.

To be sure, we possess internal clocks, beating systems of all sorts. But the world that exists beyond us tampers with these systems, leaving us with a peculiar repertoire of rhythms and tempos that dominate our lives, or the style of our lives. And events like marriage and the birth of children cause us to feel a reverberation in these rhythms, as childhood experiences, reveries, and recollections seek to overcome the connections and adjustments that we have finally settled on.

Ellie was seated at the table across from me; Ted was standing in the doorway between the dining space and living room, pacing in and out of our vision, a machine rumbling its engines, dying to lurch forward and, with a deafening rage, be gone.

"You know, Ellie," he was saying, his anger rising, "that's probably where it all goes wrong, every time with every man and every woman."

"Where?" she asked. "Where does *what* go wrong?"

"There! You see, it's right there. That you even have to ask it is what I'm saying."

"I don't see the thread of it either, Ted," I said.

"Look. You and I may come from the same background," he pointed at his wife, "but there are times when we're so different it's almost laughable that we've made it this long."

"Are the kids upstairs?" Ellie questioned.

"Oh, the hell with the kids. Let 'em hear. What the hell's the difference if they hear? What do you want, children who don't know what's going on in their own house?"

"No, I don't." Ellie's voice sounded frightened.

"That's exactly what my parents always did. They tried to keep everything from us, as though we didn't know what was going on. Let 'em hear for once."

"I'm merely. . . ."

"Look, if you'd just let me say this already."

"I'm letting you say anything you want, Ted." Ellie and I watched him walk out of the kitchen, then return.

"You've never been able to understand what my life is made of," he went on. "I tell you my plans, my prospects of what might happen at work, I tell you my dreams. That is, I used to tell you my dreams. Little that you could care."

"I listen to every one of them."

"Listening isn't enough, goddamn it. It's like Tom here. He listens. But it isn't enough. You think my plans are nuts or that I live in the future somehow. Or that I got a lot of pipe dreams."

"I didn't say that."

"You do in your way. You ought to look and listen to yourself so you could see what I see. Here's a stranger comes in this house is more sympathetic to the way I want to work out my future than the woman I married. That's really a laugh. Now that really gets me."

"That's not fair, Ted," I said.

"I don't care about fair," he shot back. "We're way beyond talking about what's fair and what's not fair." Then to me: "I tell you things about my life, you don't come back with all the junk I have to hear from her."

"*Her* is still in this room," Ellie said sharply. "Why don't you speak a little to *her?*"

"I *am* speaking to you. Who the hell you think I'm speaking to? The man in the moon?"

"I thought you might have been speaking to him." She nodded at me.

"When I want to speak with him, I'll speak with him. This is *our* business."

"Hey, maybe I should go," I suggested.

"No sir, you sit right there," Ted insisted. "You want to know about us, you listen to this part too. Every day isn't so rosy. You stay around and see the seamy side, too."

"Ted," Ellie broke in, "you're sounding foolish."

"Oh shut up!" She didn't move. "You stay, Tom. I want you to hear. I work five hard days a week. I ain't got no good prospects for anything like lots of other guys. That newspaper goes under, I'm out a job. I'd like you, Mrs. Know-it-all, to tell me just exactly what I'm supposed to do then. Huh? You got this philosophy you take every day one at a time? Isn't that your usual speech?"

"That's my usual speech," Ellie responded with resignation.

"Yeah. That's it. That's a terrific philosophy. You know where I'd be with that philosophy running my life? Do you know? *Do* you?"

"No, I don't, Ted. Where would you be?"

"You don't care. What's the use."

"I *do* care. I want to know."

"C'mon, Ted," I said, "we want to hear you."

"O.K. I'd be in the same job I had when I was eleven years old. I'd be delivering for that moron McCrackle or McCarver or whatever the hell his name was."

Ellie began to smile. "Mencken," she corrected him.

"Mencken. McCracken. What the hell's the difference? You don't understand what I'm saying anyway."

"That's not true," his wife answered him. "I do."

"Yeah? Then what am I saying?" He looked at her smugly.

"That if you hadn't had some dream or goal you wouldn't have gone beyond the life you knew as a little boy."

"That's exactly right. So what I'd like to learn from you is, if you understand that, how can you be so, so, I don't know what when it comes to listening to me?"

"I try to listen to you." Ellie's voice was kind.

"Maybe you do then. But our worlds are too different. You can't possibly understand what's in my head."

"I tell you what I believe in the best ways I can."

"That God . . . you mean that God stuff?"

"Yes," she said firmly. "That 'God stuff,' as you call it."

"That's where we part. I mean, that's where it all falls apart."

"Why? Where? What's falling apart, Ted?"

"Us. You, me. The whole thing. It falls apart. You rely on some set of beliefs that bring nothing." His anger, which had subsided, rose again. "God doesn't buy homes. God doesn't pay bills. Men do that. God doesn't provide! Love doesn't provide! You can go to church. . . ."

"I realize that."

"You can go to church every day of the week, my friend, but if I don't work, or you don't work, you don't eat. You ever hear of God throwing down food? Or clothes? Or homes? Does your friend God do *that?*"

"Of course not. That's not why people . . ."

"You see you're wrong there. I'm sorry to correct you, but that's exactly what all those poor slobs go there on Sunday for. They march off to church and give their last pennies to God and beg Him to give them food or shelter, or whatever the hell they ask for."

"I don't think that's what . . ."

"I'll tell you something else. That's exactly why they're poor."

"Who? Who's poor?" Ellie started. "What are you talking about?"

Ted bent over and pointed at her, all the while keeping his distance. "That's why the poor are poor!" he yelled at her. "Am I such a dummy that no one understands me any more? I thought I was being perfectly clear."

"Ted."

"C'mon, Ted," I joined in.

"Ted nothing," he came back. I made another gesture to leave. Ted came at me and grabbed my shoulder, pushing me back in the chair. "You want to think of me as a madman, go ahead. But you'll leave here only when I'm finished."

"C'mon, Ted," I tried again.

"No, *you* listen! Both of you. The reason the poor stay poor is because of what *you* believe," he pointed at Ellie.

"You mean *I'm* the cause of people being poor?" she asked sarcastically.

"That's exactly right. *You're* the cause. All this business of living day by day, of wanting God to solve problems, of not doing things until they actually happen, until they fall into your goddamn lap, *you* believe that, and that's why people like you don't get anywhere. People are making it in this country every day of their lives. You can still have a chance. But this business of laughing at someone 'cause he plans, or praying, that's the end. That's the living end. You hold on to your childhood; you've never let go of it; and we're married almost twenty years. You see your mother as much as you see your husband. I'd love to know what in the hell you two can find to talk about every day. That's something for you to study, Tom. If you *really* want to do an interesting study, that's what you should study: what Ellie and her mother find to talk about every day of every week of every month." Ellie had placed her head in her hands. She made no sound. "But you don't hear me complain about what portion of my salary goes to supporting her, do you?"

"No, I don't hear you complain," Ellie whispered.

"What'd you say?" Ted screamed at her.

She lifted her head and answered him more loudly: "I said, 'No, I don't hear you complain.'"

"No, I don't complain. I just shell out dough left and right for all those little things you bring her every day because that beautiful saint of a father of yours who loved everybody and everybody loved him, never planned, never dreamed. It ain't enough to be nice."

"Don't you dare speak about my father," Ellie said bitterly, staring at him.

"I'll speak about anyone I damn please. Who the hell's paying for this house anyway?"

"You don't have to speak about the dead."

"I'm not speaking about the dead. I'm speaking about poor people; people who think love and niceness and praying in a church are what matter. It's all a lot of *shit!* It's all a lot of shit! Your parents were *full* of shit!"

"Ted," she screamed at him. "Stop it!"

"And *my* parents were full of shit too. They all stunk! None of 'em knew what was going on. You know what they did? They led their lives day by day, minded their own business and decided if this is what history brought them, then that's the way it had to be. Destiny is what they call it. They just obeyed their destiny. You go ahead and look around at their children. Who's doing anything worthwhile? Huh? Who's trying to fight? Who's dreaming about the future and not about the past? Huh? You and your mother cause poverty. You sit together like mentals dreaming of the good old days when your father was alive, and none of you seems to give a rat's ass that you were poor."

"Ted, please." Ellie began to weep. My presence made it more difficult for her.

"Ted," I said, "I'm going to go."

"You're not going. I'm not done." We looked at Ellie. "Let her cry. She cries all the time. Maybe we should call her mother over and they could have a good cry together about the old days. Maybe we could make an invention and bring back the dead."

"You're a monster," Ellie sobbed.

"*I'm* the monster? Sure. Of course. *I'm* the monster, because I'm a little different from those idiots who run the church. I speak the truth. I tell you that you *waste* your life. All you have is yesterday and today. That's all all women have. Memories, tears, dreams, MOTHERS! You forget that I remember you as a child. I knew your family. They were doing nothing at all to escape their circumstances. You take a look at rich people someday. You know what they're doing? Right now? I'll tell you, and Tom will back me up, too." He continued to pace up and back across the room. "They're planning so many things for their lives they ain't got time to go to church. They got the lives of their kids planned and their kids' kids. You talk to some rich guy with his insurance policies and his trust funds. You ask them if they live day by day. They're an army, the rich. They march on the future and rip it up, eat it and spit it out. That's what they do. They beat the future at its own game. Right, Tom? They're there before it. They don't wait to see how their kids are turning out. They put 'em in the best schools, like these finishing schools, so's they can make sure the kids will *have* to turn out right."

"I don't want to hear about the rich," I heard Ellie say.

"Of course you don't. No one wants to hear what really goes on in life. You'd rather believe your daddy's going to come home and give you presents."

"I told you about talking about my father."

"O.K. *My* father then. He came home. . . . When he was alive, he didn't do one goddamn thing to change his life. Not a goddamn thing. Ever! He was supposed to be a perfectly good guy. But he was like you, all caught up with this crap of taking each day one by one and making the best of it you can. You know what you get from that philosophy?"

"Oh, stop already. Please stop."

"Ted . . ." I shouted.

"You get windows over kitchen sinks like this one." I looked at him quizzically. "Windows over kitchen sinks," he repeated, nodding his head up and down. "The rich work to make their lives work out.

Everything from their childhood on is bent toward fixing their lives and giving them security. You talk about the ability to make predictions? You should talk to the rich people like I do at work. They've got things figured out you wouldn't even be able to dream of. Believe me, I know. They think in enormous blocks of time. Not days. Not years, even. They're moving in decades, ten years, all the time left in this century. This ain't no nickel-and-dime stuff. It's all big business. Years at a time they take from the past and add onto the future. That's what their life is based on. Big business. Now, in this shitty little house which is about the best I'm ever going to be able to do, this woman here stands in front of a sink looking out the window all day worrying either about the weather, or whether it's daytime or nighttime. That's all women know. If it's daytime you make breakfast, if it's nighttime you make dinner. Bills, plans, what's going to become of us, they couldn't care less. Nighttime, daytime, all they want to know is what's in the cabinets to eat. Women and children. They're one and the same."

"So what do you want of me, Ted? What do you want of me?" Ellie's voice and tears were frightening to Ted. He moved back, disappearing into the dining area. "O.K., you're a big man, you're doing a great job here in front of us, putting on this big show, probably for Tom. You're doing fine. Now just what is it that you want? Just what do you want me and my mother and *your* children and *my* father to do? Why don't you answer that instead of carrying on here like a mental case. *You* answer, for once. What do *you* want? You want me to work? Is that it? You want the kids to stop going to school and go to work? Here, give me the phone. I'll call up the schools and pull the kids out. That's what you want, isn't it? C'mon, c'mon. Here, big man, give me the phone. Come on!" She held her open hand out in front of her. "HERE!" she screamed. Ted walked toward her.

"Stop it, goddamn it. The both of you." I heard myself yell.

"I'll give you," Ted was saying, "I'll give you the phone, across your fresh mouth is where I'll give you the phone."

"Go ahead," Ellie yelled. "You talk so much. Why don't you do something instead of babbling on like this, like some kind of an escapee from a mental hospital."

Ted's mouth closed; he nodded sarcastically. "Escapee from a mental. . . . This is what I have to take. Every day of my life. I am totally alone, making it possible for four human beings to lead their lives with a little dignity." His voice had quieted somewhat. "Four ungrateful human beings. I don't have a soul to talk to in this house. I see the way people are living. I see the way people are dying, and we're not getting any of it. Either one. I can't even afford to get us ground in the cemetery. Has that, Mrs. Big Mouth, ever crossed that brilliant brain of yours? Where, exactly, would you like them to put my corpse when I die if I don't arrange for a plot? Here? Would you like it in the kitchen maybe? Just where do the poor die? You ever think of that? That, too, is part of the future. Maybe that's what they mean by destiny. Huh? You ever think of that? I got to arrange for that too, pal.

"There's nothing for me to look forward to. I see the future that you and your mother and all those idiots you talk with at church pray to God to take care of for you. I *see* that future. I'm already seventy years old and still at the Gazette, still lifting Sunday papers, still dragging my ass around that goddamn hole. You're praying, and I'm working to have enough money to buy a place, and a way to get rid of my body which, if you'd really care to know, was dead a long, long time ago. Let me give you a little lesson." He glared at the two of us before continuing.

"Let me give you both a bit of a lesson. You," he said, pointing to me, "will have to excuse me if I don't sound like some important professor from M.I.T. A man doesn't die when his heart stops. That's not the only death a man has got to look forward to. There are lots of deaths. . . ."

"Like when your parents die?" Ellie quietly interrupted.

"NO!" he yelled at her. "Wrong again. Those are the deaths that women fall apart with. Men don't die at the sight of death. You talk to soldiers. They fight with hundreds, thousands of guys falling down all around 'em. No sir, women go to cemeteries and fall apart. Men hold women together because they know more about death than women. They know more about death because they work. They work every damn day of their lives and so they know what it is to reach a point where you can't go any further. Your own little progress is over. Dead. That's the death a man knows. The death of effort. How's that?" he questioned me. "The death of effort. You ever hear of the word incapacitated?" he asked his wife.

"Of course I did," Ellie answered.

"Of course I did," he mocked her in baby talk. "Of course you *didn't,* you mean. You can't know because women can't know the thought every day of your life of something happening and you're not able to work again. That's death, my friend." He glared at Ellie, assured that I, a man, would have to agree. "That's death."

"And women don't have that?" she inquired.

"No, women don't have that. Look at your mother. Someone else has to work so's she can eat. Someone else has to cook. Someone else has to wash up and clean and whatever you do. No sir, women don't have that. You don't know the feelings of these other deaths until you find yourself in a world where you don't have any choice; and you just stay at it, like it or not, knowing you better damn well stay healthy or a whole group of people are going to fall flat on their faces."

"You're talking through your hat," Ellie started. "How hard do you think it would be for all of us to go to work? We could make up the difference. Did *you,* Mr. Bright Ideas, ever think of *that?"*

"Many, many, many times," Ted replied softly. "Many times I've thought about what it would be like having your wife and children working while you sat around the house, sick or something. That, my friend, is another form of death. That may be the worst death of all. When women work it's a fill-in. They're a substitute, and brother, when they put in the substitutes it's because the first team either stinks or can't play. Or maybe," he smiled, "'cause the first team's got a lead they'll never catch. If only it worked that way. If only I could ever get ahead of it, instead of always chasing, chasing, chasing. . . ."

"What are you chasing, chasing, chasing?" Ellie asked, at last looking up. "Now that *is* a good question. What is it that you're always after that you can know all these *special* deaths that no one else seems to know anything about? Huh?"

"I'm chasing life!" he yelled back at her. "I'm chasing the rich and the government and my bosses and taxes and bills. That's what I'm chasing. I'm a drowning man. Look at me. I'm a drowning man, chasing after a breath. After air. Is that enough for you? I don't take life day by day and sit back and dream about the beautiful past. Anybody does that has already drowned. You chase, goddamn it. You chase your goals. You chase after the little that might possibly be coming your way. Like a fifty-dollar bonus for Christmas. *I* screw up, lady, and *you* don't get the money that *you* use to buy Christmas presents. You think bonuses just *come* to you? How do you think it works out there? That somebody just *gives* you things? Things come to you because you work your goddamn ass off, and what's more, and this no woman could ever know, you can work your ass off every day of your life, eight to six every day of your life, and end up with this!" He held his hand up, making a zero sign with his thumb and index finger. "Zero!" he yelled. "Nothing! You can work and work just like your father did and end up with nothing!"

"He was a happy man on the day he died."

"The hell he was. You don't know anything about your father. You ask your brother Milt about your father. You ask him just how happy he was in his life. He kept it from all of you. Ask your mother. She knows. Maybe now that you're a big girl she'll tell you some of the real facts of life. Go ask her when you see her tomorrow. Go tonight. Go now, for Christ's sake; it's been almost an hour since you've seen the old lady." He glanced up at the kitchen clock. I looked too and he caught me. "One more minute, then you can go too," he ordered me. "When you go there ask your mother about his drinking."

"Shut up, Ted." Ellie was angered and embarrassed. She looked at me. On several occasions we had spoken of her father's alcoholism, but the subject was not to be mentioned, apparently, when the three of us were together.

"I'm not supposed to be saying anything about that, eh? I'm supposed to keep my mouth shut so the world can think that everything around us and our beautiful childhoods was perfect? Well, I can't. You ask your mother about how many times your father died in his life before the Good Lord took him away. Ask her. She'll tell you ten thousand. Ten million. He died every night of his life. Every night he came home to that dump your family lived in, he died. You think it's easy for a man to look around every day at his life, where he works, where he lives, and be reminded of what a failure he is and always will be? That's the death part, Ellie. And that's the part, reporter, you better write about if you want to understand us. The people like us. It's that double hit; that seeing every day of your life what you got and *knowing*, knowing like your own name, that it ain't ever going to get any better. That's what you live with, and that's what your father had in his mind and in his body every day of his life. That's why a man drinks. Believe me, I know. You keep it from your children just so long. But in this country, there isn't a man who fails and holds his head up. Any man who holds his head up after he loses is just trying to protect himself. No one likes a loser. That's why the rich can't stand us. That's why we moved here. Your home, my house, what we had together all those years were *prisons*. They were advertisements of how much we didn't have. You got any doubts about it, you just turn on the T. V. and see the way people are living, with their new cars and their boats, and their homes in the country. How many times a day they have to remind you of that, I'd like to know?"

For the first time, Ellie was nodding assent. I had seen her face show relief moments before when, unthinkingly, Ted had used the phrase, "the Good Lord took him away." Now she was her husband's wife, his woman, for at last the direction of his anguish had shifted, and it was America, the rich, social classes, me, that he attacked and held responsible for the perceptions of time that he nurtured. Finally she could feel that connection with a history and with a group of people who struggle precisely because the perceptions, attitudes, and ways of others more powerful, more successful, perhaps, than they, cause them to struggle. No longer was it men against women; it was the poor against the rich, those who can, with a genuine sense of autonomy and possibility underwriting it, achieve with dignity, and those who must sweat out the work, the waiting, and the decisions of bosses for bonuses and a chance to move ahead slightly.

Ted stood, not exactly next to his wife, but closer to her, leaving me, still seated, the third person. Reflecting on his words, I sought a proper way to excuse myself. For minutes the three of us were silent, Ellie and I staring at the table top, Ted standing near us, motionless. I could hear him breathing, and I could feel my eyebrows rising and falling, as though something in me wanted to signal the two of them. We will all three of us go on from here, is what the silence meant to me. We will all face the forms of death, as Ted had said, that each of us faces, not every day perhaps, but often enough. And we will dream of a past, of homes perhaps where it is warm, and where everyone is content. We will know experiences that make us live moment by moment, for as long as each moment lasts, just as we will think up something for the future, a plan, a prediction, a style of ignoring it, almost as children generate some new idea for a game they have played before.

Ted broke the silence. His voice was even and resolute. "I want you to know, Tom, that you have my permission to write anything you want about this afternoon. All of it. I'm pretty sure Ellie feels the same way."

"I do. Yes, I do," Ellie said.

When I looked up at them they were already looking intently at me. "I appreciate that," I responded. "I don't ever know exactly what I'll be writing, of course."

"I don't know much about writing," Ted began. "I don't want to ever tell a man how to do his work if I don't know how to do that work better than that man. But I think it might just be good for you to take a shot at writing this. Don't worry about us. You go ahead and write what you feel you should. Call us about our meeting next week." That was all he said that afternoon.

Two feelings had come over me, two feelings as distinct as shaking hands with both Grazianos simultaneously, Ted's right hand in my right hand, Ellie's hand in my left hand. I felt, first, that I had been as moved by these two people as I had ever been by anyone or anything, and that because of this resonance deep inside me, our lives were now married. But I felt, too, a desire to make all three of us, through my writing, special and unforgettable; famous, I suppose. I was half-way home, driving through an angry rain that made visibility almost impossible, before the one word that swirled in my head finally came to rest: immortality.

In the end, the overriding dimension of time in human experience is the dimension of freedom. For freedom implies the right to embrace one's past, one's immediate experiences, and one's prospects as one chooses, all according to an equitable reality. Like the word "future," freedom implies possibility. When freedom is denied, therefore, a person's past, present, or future, or all three, may also be denied. And so a past is rejected, prospects are ruled outrageous, and a man and woman remain alone to suffer an anguish they are led to believe each has caused the other. In fact, it is a culture's power and the ways of social systems that sting human possibility and kill the belief in an honorable past, a tolerable present, and the prospect of a generous future.

Appendix:
Note on the Methodology
of Life Studies

For several years I have been doing research in various Boston communities. The research has been devoted to describing how people in these communities live their lives. Some of the communities are known for their great wealth, others for their extreme poverty, whereas still others are what we normally think of as being properly middle class. My work involves meeting with people, talking about their lives, and in some instances writing about them, and, in a sense, about them and me together.* To a certain degree, I have become attracted to people who have, in one way or another, been hurt, either personally through some act of fate or misfortune, or by society in the form of what we call economic hardship or social or political disenfranchisement. These people survive—in not an especially courageous manner as they see it—and continue to believe in the goodness of life.

It is not surprising that the experience of time should emerge as a central theme in all of these lives. What, after all, is the study of a life, if not the study of time in human experience? All people work out propositions about time and fashion theoretical schemes that they hope will yield some sense of order and meaning for what they call personal development, destiny, or achievement. Still, as this book developed, I began to listen for more explicit expressions about time

*In this regard see Cottle, 1974.

257

from those with whom I visited. But there was nothing unusual to listen for. All along, apparently, our conversations had been about time in human experience, and about the ways in which these people manage not formidable philosophical concepts exactly, but philosophies that encompass their lives. The concepts, visions, and perceptions I had resolved to listen for were already there, not simply displayed, but also not embedded in language that only experts can comprehend.

The research of the type the life studies in this volume represent are predicated on observation, description, and on allowing people to speak for themselves. Explicit analysis, or interpretation of their language and actions, plays a far less significant role in the life study than in what we usually think of as a clinical case study. But the term should not hide the psychological and sociological propositions and findings that in fact underlie the research.

It was purely by accident that I came to know the people represented in the life studies. I had been speaking with young war veterans in a Boston hospital when I met Marcus Nathaniel Simpson (Chapter 2), who was there recovering from a series of illnesses. Gradually, I spent more and more time with him, listening to his own personal ruminations on, and overview of, time and human experience. In the case of Matilda Rutherford (Chapter 5), our meeting was arranged by a mutual friend. One afternoon they asked me to join them at a small neighborhood restaurant.

Meeting Mr. and Mrs. Graziano (Chapter 8) was also a pleasant accident. I had been following the educational career of a young man when he introduced me to his employer. It happened to be Ted Graziano. Soon afterward I met Ted's wife, Eleanor, their children, and other relatives of theirs as well. As the Grazianos were among the first families I had chosen to "research," in the beginning I hardly knew what to ask them. In truth, during our early meetings they interviewed me, and on several occasions, Mr. Graziano threatened to write up our first meetings in order to depict what he calls "the birth of a census taker."

My method in this work is to meet with people as often as our schedules allow. Always it is at their homes, places of work or schools, somewhere that is comfortable for them. We may meet as often as several times a week or as infrequently as once a month. Meetings normally last an hour or two, although sometimes they take up a whole afternoon. When situations prevent us from getting together, a telephone call helps to exchange lingering ideas and feelings. As a rule, I never write about anyone before knowing him at least a year. I have no set policy about this; but from the standpoint of friendship, the collection of materials, and ethics, it does not seem proper to begin describing a life any earlier. There are, of course, exceptions to this rule, as in the case of Marcus Simpson, whose association with me lasted less than four months. Normally, however, after about two years with someone, a phase seems to be completed and writing becomes a more logical next step.

When I began this research, I used a tape recorder. Gradually, however, I grew uneasy with the device, the style of interaction it partially evoked, and even more, the technique it demanded. So I did away with it, and used it instead for dictating dialogues and ideas immediately after conversations had been concluded. Without the machine, I have surely lost a sizable amount of information, but I also have gained some advantages. For one thing, I am comfortable and feel less foolish than I did when thrusting a microphone in someone's face. For another thing, the absence of the tape recorder forced me to rely on myself, and more specifically, on what the conversation evoked in me. It forced me, moreover, to listen and to remember, and to wonder about what I might be forgetting or missing. It also forced me to confront my own feelings and become implicated in the lives of these people. No longer could I hide behind a mechanical device and the technique that this device implied. With the tape recorder removed, I felt freer to receive their words, as well as freer to talk myself, and I began to participate in a friendship rather than to conduct an interview.

In all cases I have changed the names of the people included in the life studies and enough of the facts of their lives to preserve confidentiality. They have all checked the validity of the conversations by reading the manuscripts. As a rule, I ask them first to confirm the information and style. Next, and equally important, I ask them whether they can recognize themselves and our relationship in the manuscript. When they cannot, and this happens even when we both believe that the information in the text is accurate, I begin again with another draft—until they approve. On a few occasions I have felt what a portrait artist must feel when he has difficulty "capturing" a person. On these occasions I have put the manuscript aside, and while I continue to see those persons, it may be that my work with them will never evolve into a manuscript. Something about them, and even more, something about me, makes the task of writing impossible.

Regularly, I am advised to take for granted that everyone understands the complexity of human nature and to speak only to one or two issues that readers might find enlightening. No one needs to be told again of the richness of a single life, I have been counseled—that we all know. As I reflect, however, on the people with whom I have spoken over the last years, I cannot so quickly shed the urge to insist on the issue of behavioral complexity and on human beings' capacity to articulate this complexity. More precisely, I am struck by people's ability to speak about and comprehend the meaning of their past, present, and future. In human expression, in anger and despair, in hopes, regrets, prospects, and resiliency lie the interpretations of human behavior and the theories that all of us construct and test out in our lives.

The sorts of theoretical and empirical questions that we have considered in this book linger with me during my conversations in the communities where my research takes me. They haunt me, as does a young woman's polite response to one of my so-called therapeutic interpretations: "Yes, that's part of it. That's part of it—but not all." Nonetheless, the methods and findings of the inquiries examined in Chapters 1, 3, 4, and 6 filled my head even as I spoke with Marcus Simpson, Matilda Rutherford, and Theodore and Eleanor Graziano. As much as I sought to exclude this other kind of research, it returned as if some inner voice insisted that I recognize that my observational and descriptive work was not so distant, after all, from this more impersonal research into time.

It is not surprising that the people represented in the life studies should be capable of making such realistic and poetic images of time. Their lives remind us that the mysteries of time are made comprehensible through language and expression. We need not, however, go beyond their accounts, their single lives, but should approach them as one might edge closer to a painting, or to a piece of sculpture. We seek to touch art, and thereby to touch the life of the artist. Similarly, we wish to be touched by the artist. In a way, the impersonal methods of inquiry into perceptions of time that we have just reviewed provide information that ultimately enhances the likelihood of our being touched by another person. When our prior and prospective intellectual and emotional worlds are brought forth, we are more likely to know another person, and to feel, through this conjoining of experiences, the touch of something as genuine and still metaphysical as time.

—T. J. C.

References

Aberle, D. F., and Naegele, K. D. 1952. Middle-class fathers' occupational role and attitudes toward children. *American Journal of Orthopsychiatry* 22:366-78.

Allen, V. L. 1970. The psychology of poverty: problems and prospects. In *Psychological factors in poverty,* ed. V. L. Allen, pp. 367-83. Chicago: Markham.

Allman, J. M. 1972. Education and social mobility in Tunisia. Ph.D. Dissertation, Harvard University.

Allport, G. W. 1961. *Pattern and growth in personality.* New York: Holt, Rinehart & Winston.

Ames, L. B. 1946. The development of the time sense in the young child. *Journal of Genetic Psychology* 68:97-125.

Arieti, S. 1947. The processes of expectation and anticipation. *Journal of Nervous and Mental Diseases* 106:471-81.

Armer, M., and Youtz, R. 1971. Formal education and individual modernity in an African society. *American Journal of Sociology* 76:604-26.

Back, K. W., and Gergen, K. J. 1963. Apocalyptic and serial time orientations and the structure of opinions. *Public Opinion Quarterly* 27:427-42.

Baillie, J. 1951. *The belief in progress.* New York: Scribner.

Bakan, D. 1966. *The duality of human existence.* Chicago: Rand McNally.

Balandier, G. 1954. Comparative study of economic motivations and incentives in a traditional and in a modern environment. *International Social Science Bulletin* 6:372-87.

Banfield, E. C. 1968. *The unheavenly city: the nature and future of our urban crisis.* Boston: Little, Brown.

Barker, R. B., Dembo, T., and Lewin, K. 1943. Frustration and regression. In *Child behavior and development,* eds. R. G. Barker, J. S. Kounin, and H. F. Wright, pp. 441-58. New York: McGraw-Hill.

Barry, H., Child, I. L., and Bacon, M. K. 1959. Relation of child training to subsistence economy. *American Anthropologist* 61:51-63.

Battle, E. S., and Rotter, J. B. 1963. Children's feelings of personal control as related to social class and ethnic group. *Journal of Personality* 31:482-90.

Beilin, H. 1956. The pattern of postponability and its relation to social class mobility. *Journal of Social Psychology* 44:33-48.

Bell, W., and Mau, J. A. 1971. Images of the future: theory and research strategies. In *The sociology of the future,* eds. W. Bell and J. A. Mau, pp. 6-44. New York: Russell Sage.

Berger, G. 1967. A phenomenological approach to the problem of time. In *Readings in existential phenomenology,* eds. N. Lawrence and D. O'Connor, pp. 148-204. Englewood Cliffs: Prentice-Hall.

Berger, P. L. 1963. *Invitation to sociology: a humanistic perspective.* Garden City: Doubleday.

Bernot, L., and Blancard, R. 1953. *Nouville: un village français.* Paris: Institut d'Ethnologie.

Bender, L. 1950. Anxiety in disturbed children. In *Anxiety,* eds. P. H. Hoch and J. Zubin, pp. 119-39. New York: Grune & Stratton.

Bettelheim, B. 1958. Individual and mass behavior in extreme situations. In *Readings in social psychology.* 3rd ed., eds. E. E. Maccoby, T. M. Newcomb, and E. L. Hartley, pp. 300-310. New York: Holt.

Binswanger, L. 1958. The case of Ellen West; an anthropological-clinical study. In *Existence: a new dimension in psychiatry and psychology,* eds. R. May, E. Angel, and H. F. Ellenberger, pp. 37-91. New York: Basic Books.

Bleuler, E. (1912). Autistic thinking. In *Organization and pathology of thought,* ed. D. Rapaport, pp. 399-437. New York: Columbia University Press, 1951.

Blos, P. 1962. *On adolescence: a psychoanalytic interpretation.* Glencoe: Free Press.

Bochner, S., and David, K. H. 1968. Delay of gratification, age and intelligence in an aboriginal culture. *International Journal of Psychology* 3, no. 3:167-74.

Bohannan, P. 1953. Concepts of time among the Tiv of Nigeria. *Southwestern Journal of Anthropology* 9:251-62.

Bourdieu, P. 1963. The attitude of the Algerian peasant toward time. In *Mediterranean countrymen,* ed. J. Pitt-Rivers, pp. 55-72. The Hague: Mouton.

Bowlby, J. 1951. *Maternal care and mental health.* World Health Organization Monograph Series, Whole No. 2.

Brim, O. G., Jr., and Forer, R. A. 1956. A note on the relation of values and social structure to life planning. *Sociometry* 19:54-60.

Brown, L. C. 1965. Tunisia. In *Education and political development,* ed. J. S. Coleman, pp. 144-68. Princeton: Princeton University Press.

264   References

Brown, R. 1965. *Social psychology.* New York: Free Press.

Bruner, J. S. 1960. *The process of education.* New York: Vintage.

Bugental, J., ed. 1967. *Challenges of humanistic psychology.* New York: McGraw-Hill.

Bühler, C. 1968. The integrating self. In *The course of human life,* eds. C. Bühler and F. Massarik, pp. 330–50. New York: Springer.

Burton, R. V., and Whiting, J. W. M. 1961. The absent father and cross-sex identity. *Merrill-Palmer Quarterly of Behavior and Development* 7:85–95.

Bury, J. B. 1932. *The idea of progress: an inquiry into its growth and origin.* New York: Dover.

Camilleri, C. 1970. Les attitudes et représentations familiales des jeunes dans un pays décolonisé en voie de développement: essai sur le changement socio-culturel dans un pays du tiers-monde (Tunisie). Doctorat-ès-lettres, University of Paris.

Campbell, E. Q. 1969. Adolescent socialization. In *Handbook of socialization theory and research,* ed. D. A. Goslin, pp. 821–59. Chicago: Rand McNally.

Campbell, J. 1954. Functional organization of the central nervous system with respect to orientation in time. *Neurology* 4:295–300.

Carr, E. H. 1961. *What is history?* New York: Knopf.

Cleckley, H. 1959. Psychopathic states. In *American handbook of psychiatry.* vol. I, ed. S. Arieti, pp. 567–88. New York: Basic Books.

Cohen, A. K. 1955. *Delinquent boys.* New York: Free Press.

Cohen, J. 1964. Psychological time. *Scientific American* 211:116–24.

———. 1966. Subjective time. In *The voices of time,* ed. J. T. Fraser, pp. 257–75. New York: Braziller.

Cope, L. 1919. Calendars of the Indians north of Mexico. *University of California Publications in American Archeology and Ethnology* 16:119–76.

Coser, L. A., and Coser, R. L. 1963. Time perspective and social structure. In *Modern sociology: an introduction to the study of human interaction,* A. W. Gouldner and H. P. Gouldner, pp. 638–47. New York: Harcourt, Brace & World.

Cottle, T. J. 1967. The circles test: an investigation of temporal relatedness and dominance. *Journal of Projective Techniques and Personality Assessment* 31:58–71.

———. 1968a. The duration inventory: subjective extensions of temporal zones. *Acta Psychologica* 29:1–20.

———. 1968b. The location of experience: a manifest time orientation. *Acta Psychologica* 28:129–149.

———. 1969a. Future orientations and avoidance: speculations on the time of achievement and social roles. *Sociological Quarterly* 10:419–37.

———. 1969b. The money game: notes on fantasies of temporal recovery and preknowledge. *Diogenes* 65:110–34.

———. 1969c. Temporal correlates of the achievement value and manifest anxiety. *Journal of Consulting and Clinical Psychology* 33:541–50.

———. 1973. *The abandoners: portraits of loss, separation and neglect.* Boston: Little, Brown.

———. *Black children, white dreams.* Boston: Houghton Mifflin, 1974.

Cottle, T. J., and Howard, P. 1969. Temporal extension and time zone bracketing in Indian adolescents. *Journal of Perceptual and Motor Skills* 28:599–612.

Davids, A., and Parenti, A. N. 1958. Time orientation and interpersonal relations of emotionally disturbed and normal children. *Journal of Abnormal and Social Psychology* 57:299–305.

Davis, A. 1946. The motivation of the underprivileged worker. In *Industry and society,* ed. W. F. Whyte, pp. 84–106. New York: McGraw-Hill.

Demeerseman, A. 1967. *La famille tunisienne et les temps nouveaux: essai de psychologie sociale.* Tunis: Maison Tunisienne de l'Edition.

Dollard, J., and Miller, N. E. 1950. *Personality and psychotherapy: an analysis in terms of learning, thinking, and culture.* New York: McGraw-Hill.

Doob, L. W. 1960. *Becoming more civilized: a psychological exploration.* New Haven: Yale University Press.

Douvan, E., and Adelson, J. 1958. The psychodynamics of social mobility in adolescent boys. *Journal of Abnormal and Social Psychology* 56: 31–44.

———. 1966. *The adolescent experience.* New York: Wiley.

Dummett, M. 1967. Bringing about the past. In *The philosophy of time,* ed. R. M. Gale, pp. 252–74. Garden City: Doubleday.

Eisenberg, P., and Lazarsfeld, P. 1938. The psychological effects of unemployment. *Psychological Bulletin* 35:358–90.

Epley, D., and Ricks, D. R. 1963. Foresight and hindsight on the TAT. *Journal of Projective Techniques* 27:51–59.

Erasmus, C. J. 1961. *Man takes control: cultural development and American aid.* Minneapolis: University of Minnesota Press.

Erikson, E. H. 1950. *Childhood and society.* New York: Norton.

———. 1959. Identity and the life cycle: selected papers. *Psychological Issues* 1:Whole No. 1.

———. 1964a. *Insight and responsibility: lectures on the ethical implications of psychoanalytic insight.* New York: Norton.

———. 1964b. Inner and outer space: reflections on womanhood. *Daedalus* 93:582–606.

———. 1968. *Identity: Youth and Crisis.* New York: Norton.

Eson, M. E., and Greenfield, S. 1962. Life space: its content and temporal dimensions. *Journal of Genetic Psychology* 100:113–28.

Evans-Pritchard, E. E. 1940. *The Nuer.* Oxford: Clarendon.

Ezekiel, R. S. 1968. The personal future and peace corps competence. *Journal of Personality and Social Psychology Monograph Supplement* 8:No. 2, part 2.

Farber, M. L. 1944. Suffering and time-perspective of the prisoner. *University of Iowa Studies in Child Welfare* 20:153–227.

———. 1951. The armageddon complex: dynamics of opinion. *Public Opinion Quarterly* 15:217–24.

———. 1953. Time perspective and feeling tone: a study in the perception of the days. *Journal of Psychology* 35:253–57.

Farnham-Diggory, S. 1966. Self, future and time: a developmental study of the concepts of psychotic, brain-damaged and normal children. *Monograph of the Society for Research in Child Development* 31: Whole No. 1.

Finestone, H. 1957. Cats, kicks, and color. *Social Problems* 5:3–13.

Flavell, J. H. 1963. *The developmental psychology of Jean Piaget.* Princeton: Van Nostrand.

Foster, G. M. 1962. *Traditional cultures: and the impact of technological change.* New York: Harper & Row.

Foster, P. 1965. *Education and social change in Ghana.* Chicago: University of Chicago Press.

Fraisse, P. 1963. *The psychology of time.* Trans. by J. Leith. New York: Harper & Row.

Frank, L. K. 1950. Time perspectives. In *Society as patient: essays on culture and personality,* pp. 339–58. New Brunswick: Rutgers University Press.

Freeman, W., and Watts, J. W. 1942. *Psychosurgery: intelligence, emotion and social behavior following prefrontal lobotomy for mental disorders.* Springfield: Thomas.

Freud, S. (1911). Formulations regarding the two principles in mental functioning. In *Collected papers,* vol. IV, pp. 13–21. New York: Basic Books, 1959.

Friedmann, F. G. 1953. The world of "La Miseria." *Partisan Review* 20:218–31.

Gans, H. J. 1962. *The urban villagers: group and class in the life of Italian-Americans.* New York: Free Press.

Geertz, C. 1966. *Person, time, and conduct in Bali: an essay in cultural analysis.* Southeast Asia Studies, Cultural Report Series No. 14. New Haven: Yale University Press.

Ginzberg, E., Ginsburg, S. W., Axelrad, S., and Herma, J. L. 1951. *Occupational choice: an approach to a general theory.* New York: Columbia University Press.

Goffman, E. 1961. *Asylums.* Chicago: Aldine.

Goldfarb, W. 1945. Psychological privation in infancy and subsequent adjustment. *American Journal of Orthopsychiatry* 15:249-55.

Graves, T. D. 1961. Time perspective and the deferred gratification pattern in a tri-ethnic community. Ph.D. Dissertation, University of Pennsylvania.

Gurvitch, G. 1964. *The spectrum of social time.* Dordrecht: Reidel.

Hallowell, A. I. 1955. *Culture and experience.* Philadelphia: University of Pennsylvania Press.

Hare, R. D. 1965. Psychopathy, fear arousal and anticipated pain. *Psychological Reports* 16:499-502.

Hartley, R. E. 1959. Sex-role pressures and the socialization of the male child. *Psychological Reports* 5:457-68.

Hearnshaw, L. W. 1956. Temporal integration and behavior. *Bulletin of the British Psychological Society* 30:1-20.

Hebb, D. O. 1966. *A textbook of psychology.* 2nd ed. Philadelphia: Saunders.

Hebb, D. O., and Thompson, W. R. 1968. The social significance of animal studies. In *The handbook of social psychology,* vol. II. 2nd ed., eds. G. Lindzey and E. Aronson, pp. 729-74. Reading: Addison-Wesley.

Heilbroner, R. L. 1959. *The future as history.* New York: Grove Press.

———. 1963. *The great ascent: the struggle for economic development in our time.* New York: Harper & Row.

Henry, W. E. 1949. The business executive: the psychodynamics of a social role. *American Journal of Sociology* 54:286-91.

Hess, E. H. 1962. Ethology: an approach toward the complete analysis of behavior. In *New directions in psychology,* R. Brown, E. Galanter, E. H. Hess, and G. Mandler, pp. 157-266. New York: Holt, Rinehart & Winston.

Hoffer, E. 1951. *The true believer: thoughts on the nature of mass movements.* New York: New American Library.

Horowitz, I. L. 1972. *Three worlds of development: the theory and practice of international stratification.* 2nd ed. New York: Oxford University Press.

Hughes, C. G., Tremblay, M., Rapoport, R. N., and Leighton, A. H. 1960. *People of cove and woodlot: communities from the viewpoint of social psychiatry.* New York: Basic Books.

Illich, I. 1971. *Deschooling society.* New York: Harper & Row.

Inhelder, B., and Piaget, J. 1958. *The growth of logical thinking from childhood to adolescence.* Trans. by A. Parsons and S. Milgram. New York: Basic Books.

Inkeles, A. 1955. Social change and social character: the role of parental mediation. *Journal of Social Issues* 11:12-23.

———. 1966. The modernization of man. In *Modernization: the dynamics of growth,* ed. M. Weiner, pp. 138-50. New York: Basic Books.

———. 1969. Making men modern: on the causes and consequences of individual change in six developing countries. *American Journal of Sociology* 75:208-25.

Israeli, N. 1936. *Abnormal personality and time.* New York: Science Press.

Jacobsen, C. F. 1936. Studies of cerebral function in primates. *Comparative Psychology Monographs* 13:1-60.

Jacques, E. 1956. *Measurement of responsibility: a study of work, payment and individual capacity.* London: Tavistock.

Jahoda, G. 1968. Some research problems in African education. *Journal of Social Issues* 24:161-75.

Jones, R. E. 1949. Personality changes in psychotics following prefrontal lobotomy. *Journal of Abnormal and Social Psychology* 44:315-28.

Kagan, J. 1964. Acquisition and significance of sex typing and sex role identity. In *Review of child development research.* vol. I, eds. M. L. Hoffman and L. W. Hoffman, pp. 137-67. New York: Russell Sage.

Kahl, J. A. 1962. *The American class structure.* New York: Holt, Rinehart & Winston.

———. 1968. *The measurement of modernism: a study of values in Brazil and Mexico.* Austin: University of Texas Press.

Kelly, G. A. 1958. Man's construction of his alternatives. In *Assessment of human motives,* ed. G. Lindzey, pp. 33-64. New York: Grove Press.

Keniston, K. 1965. *The uncommitted: alienated youth in American society.* New York: Dell.

Kinsey, A. C., Pomeroy, W. B., and Martin, C. E. 1948. *Sexual behavior in the human male.* Philadelphia: Saunders.

Klineberg, S. L. 1967. Changes in outlook on the future between childhood and adolescence. *Journal of Personality and Social Psychology* 7:185-93.

———. 1968. Future time perspective and the preference for delayed reward. *Journal of Personality and Social Psychology* 8:253-57.

———. 1971. Modernization and the adolescent experience: a study in Tunisia. *The Key Reporter* 37, No. 1:2-4.

———. 1972. The relative contribution of parents and schools to the modernization of adolescents. Proceedings of the 67th Annual Meeting of the American Sociological Association.

Kluckhohn, F., and Strodtbeck, F. L. 1961. *Variations in value orientations.* Evanston: Row, Peterson.

Knapp, R. H., and Garbutt, J. T. 1958. Time imagery and the achievement motive. *Journal of Personality* 26:426-34.

———. 1965. Variation in time descriptions and in achievement. *Journal of Social Psychology* 67:269-72.

Köhler, W. 1925. *The mentality of apes.* New York: Vintage.

Krauss, H. H. 1967. Anxiety: the dread of a future event. *Journal of Individual Psychology* 23:88-93.

Krauss, H. H., and Ruiz, R. A. 1967. Anxiety and temporal perspective. *Journal of Clinical Psychology* 23:340-42.

Leach, E. R. 1961. *Rethinking anthropology.* London: Athalone.

Lee, D. 1959. *Freedom and culture.* Englewood Cliffs: Prentice-Hall.

Lehman, H. C., and Witty, P. A. 1931. A study of vocational attitudes in relation to pubescence. *American Journal of Psychology* 43:93-101.

Lerner, D. 1958. *The passing of traditional society: modernizing the Middle East.* Glencoe: Free Press.

———. 1963. Toward a communication theory of modernization: a set of considerations. In *Communications and political development,* ed. L. W. Pye, pp. 327-50. Princeton: Princeton University Press.

Lessing, E. E. 1968. Demographic, developmental, and personality correlates of length of future time perspective (FTP). *Journal of Personality* 36:183-201.

Lévi-Strauss, C. 1960. The problem of invariance in anthropology. *Diogenes* 31:19-28.

———. 1966. *The savage mind.* Chicago: University of Chicago Press.

Lévy-Valensi, E. A. 1965. *Le temps dans la vie psychologique.* Paris: Flammarion.

Lewin, K. 1935. *A dynamic theory of personality.* Trans. by D. K. Adams and K. E. Zener. New York: McGraw-Hill.

————. 1951. *Field theory in social science,* ed. D. Cartwright. New York: Harper.

Lewis, O. 1961. *The children of Sanchez.* New York: Random House.

————. 1966a. *La vida; a Puerto Rican family in the culture of poverty—San Juan and New York.* New York: Random House.

————. 1966b. The culture of poverty. *Scientific American* 215:19-25.

————. 1969. Review of C. A. Valentine, *Culture and Poverty. Current Anthropology* 10:2-3.

Liebow, E. 1967. *Tally's corner: a study of Negro streetcorner men.* Boston: Little, Brown.

Lifton, R. J. 1964. Individual patterns in historical change: imagery of Japanese youth. *Comparative Studies in Society and History* 6:369-383.

Lindner, R. M. 1944. *Rebel without a cause: the story of a criminal psychopath.* New York: Grune & Stratton.

Lipman, A., and Havens, A. E. 1965. The Columbia Violencia: an ex post facto experiment. *Social Forces* 44:238-45.

Lykken, D. T. 1957. A study of anxiety in the sociopathic personality. *Journal of Abnormal and Social Psychology* 55:6-10.

Maccoby, E. ed. 1966. *The development of sex differences.* Stanford: Stanford University Press.

MacIver, R. M. 1962. *The challenge of the passing years: my encounter with time.* New York: Simon & Schuster.

MacKinnon, D. W. 1944. A topological analysis of anxiety. *Character and Personality* 12:163-76.

Malinowski, B. 1935. *Coral gardens and their magic.* New York: American Book Co.

Malmo, R. B. 1942. Interference in delayed response in monkeys after removal of the frontal lobes. *Journal of Neurophysiology* 5:295-308.

Marquis, D. P. 1941. Learning in the neonate. *Journal of Experimental Psychology* 29:263-82.

Maslow, A. H. 1962. *Toward a psychology of being.* Princeton: Van Nostrand.

May, R. 1953. *Man's search for himself.* New York: Norton.

————. 1958. Contributions of existential psychotherapy. In *Existence,* eds. R. May, E. Angel, and H. F. Ellenberger, pp. 37-91. New York: Basic Books.

McClelland, D. C. 1961. *The achieving society.* Princeton: Van Nostrand.

McCord, W., and McCord, J. 1964. *The psychopath: an essay on the criminal mind.* Princeton: Van Nostrand.

McDougall, W. 1923. *Outline of psychology.* New York: Scribner.

McQueen, A. J. 1968. Education and marginality of African youth. *Journal of Social Issues* 24:179-94.

Mead, G. H. (1932). Time. In *George Herbert Mead on social psychology,* ed. A. Strauss, pp. 328-41. Chicago: University of Chicago Press, 1956.

Mead, M. 1949. *Male and female.* New York: Morrow.

———. ed. 1955. *Cultural patterns and technical change.* New York: New American Library.

———. 1956. *New lives for old: cultural transformation, Manus, 1928-1953.* New York: New American Library.

———. 1970. *Culture and commitment: a study of the generation gap.* Garden City: Doubleday.

Meerloo, J. A. M. 1954. *The two faces of man: two studies on the sense of time and on ambivalence.* New York: International Universities Press.

Meier, D. L., and Bell, W. 1959. Anomia and differential access to the achievement of life goals. *American Sociological Review* 24:189-202.

Melikan, L. 1959. Preference for delayed reinforcement: an experimental study among Palestinian Arab refugee children. *Journal of Social Psychology* 50:81-86.

Micaud, C. A. 1964. Social and economic change. In *Tunisia: the politics of modernization,* C. A. Micaud, L. C. Brown, and C. H. Moore, pp. 131-90. New York: Praeger.

Michaud, E. 1949. *Essai sur l'organisation de la connaissance entre 10 et 14 ans.* Paris: Vrin.

Miller, S. M., Riessman, F., and Seagull, A. A. 1965. Poverty and self-indulgence: a critique of the non-deferred gratification pattern. In *Poverty in America,* eds. L. A. Ferman, J. L. Kornbluh, and A. Haber, pp. 285-302. Ann Arbor: University of Michigan Press.

Minkowski, E. 1958. Findings in a case of schizophrenic depression. In *Existence,* eds. R. May, E. Angel, and H. F. Ellenberger, pp. 127-38. New York: Basic Books.

Mischel, W. 1958. Preference for delayed reinforcement: an experimental study of a cultural observation. *Journal of Abnormal and Social Psychology* 56:57-61.

———. 1961a. Preference for delayed reinforcement and social responsibility. *Journal of Abnormal and Social Psychology* 62:1-7.

———. 1961b. Father absence and delay of gratification: cross-cultural comparisons. *Journal of Abnormal and Social Psychology* 63:116-24.

———. 1970. Sex-typing and socialization. In *Carmichael's manual of child psychology.* 3rd ed., ed. P. H. Mussen, pp. 3-72. New York: Wiley.

Mischel, W., and Metzner, R. 1962. Preference for delayed reward as a function of age, intelligence, and length of delay interval. *Journal of Abnormal and Social Psychology* 64:425-31.

Moore, C. H. 1964. The era of the Neo-Destour. In *Tunisia: the politics of modernization*, C. A. Micaud, L. C. Brown, and C. H. Moore, pp. 69-128. New York: Praeger.

Moore, W. E. 1963. *Man, time, and society*. New York: Wiley.

Mowrer, O. H., and Ullman, A. D. 1945. Time as a determinant in integrative learning. *Psychological Review* 52:61-90.

Murray, H. A. 1959. Preparations for the scaffold of a comprehensive system. In *Psychology: a study of a science*, vol. III, ed. S. Koch, pp. 7-54. New York: McGraw-Hill.

Mussen, P. H., Conger, J. J., and Kagan, J. 1963. *Child development and personality*. 2nd ed. New York: Harper & Row.

Niehoff, A. H. 1966a. *A casebook of social change*. Chicago: Aldine.

————, ed. 1966b. Fatalism in Asia: old myths and new realities. *Anthropological Quarterly* 39: Whole No. 3.

Osgood, C. E. 1962. *An alternative to war or surrender*. Urbana: University of Illinois Press.

Parsons, T., and Bales, R. F. 1955. *Family, socialization and interaction process*. Glencoe: Free Press.

Petrie, A. 1952. *Personality and the frontal lobes: an investigation of the psychological effects of different types of leucotomy*. New York: Blakiston.

Piaget, J. 1952. *The origins of intelligence in children*. Trans. by M. Cook. New York: International Universities Press.

————. 1955. The development of time concepts in the child. In *Psychopathology of childhood*, eds. P. Hoch and J. Zubin, pp. 34-44. New York: Grune & Stratton.

————. 1966. Time perception in children. In *The voices of time*, ed. J. T. Fraser, pp. 31-55. New York: Braziller.

Pleck, J. H. 1973. Psychological frontiers for men. Paper presented at Brandeis University, February 22.

Rabin, A. I. 1965. *Growing up in the Kibbutz*. New York: Springer.

Redl, F., and Wineman, D. 1951. *Children who hate*. New York: Collier.

Roberts, A. H., and Herrmann, R. S. 1960. Dogmatism, time perspective, and anomie. *Journal of Individual Psychology* 16:67-72.

Roby, T. B. 1962. Utility and futurity. *Behavioral Science* 7:194-210.

Rodgers, W. B. 1967. Changing gratification orientations: some findings from the Out Island Bahamas. *Human Organization* 26:200-205.

Rokeach, M. 1960. *The open and the closed mind: investigations into the nature of belief systems and personality systems.* New York: Basic Books.

Rokeach, M., and Bonier, R. 1960. Time perspective, dogmatism, and anxiety. In *The open and the closed mind,* M. Rokeach, pp. 366–75. New York: Basic Books.

Rosen, B. 1956. The achievement syndrome: a psychocultural dimension of social stratification. *American Sociological Review* 21:203–11.

Roth, J. A. 1963. *Timetables.* Indianapolis: Bobbs-Merrill.

Rotter, J. B. 1966. Generalized expectancies for internal vs. external control of reinforcement *Psychological Monographs* 80:Whole No. 609.

Rubin, V., and Zavalloni, M. 1969. *We wish to be looked upon: a study of the aspirations of youth in a developing society.* New York: Teachers College Press.

Ruiz, R. A., and Krauss, H. H. 1968. Anxiety, temporal perspective, and item content of the incomplete thoughts test (ITT). *Journal of Clinical Psychology* 24:70–72.

Sattler, J. M. 1964. Counselor competence, interest, and time perspective. *Journal of Counseling Psychology* 4:357–60.

Schafer, R. 1958. Regression in the service of the ego: the relevance of a psychoanalytic concept for personality assessment. In *Assessment of human motives,* ed. G. Lindzey, pp. 119–48. New York: Grove Press.

Schnaiberg, A. 1970. Measuring modernism: theoretical and empirical explorations. *American Journal of Sociology* 76:399–425.

Schneider, L., and Lysgaard, S. 1953. The deferred gratification pattern: a preliminary study. *American Sociological Review* 18:142–49.

Schramm, W. 1964. *Mass media and national development: the role of information in developing countries.* Stanford: Stanford University Press.

Seagull, A. A. 1964. The ability to delay gratification. Ph.D. Dissertation, Syracuse University.

Seeley, J. R., Sim, R. A., and Loosely, E. W. 1956. *Crestwood Heights: a study of the culture of suburban life.* New York: Basic Books.

Shackle, G. L. S. 1958. *Time in economics.* Amsterdam: North-Holland.

Shybut, J. 1963. Delayed gratification: a study of its measurement and its relationship to certain behavioral, psychological and demographical variables. Master's Thesis, University of Colorado.

Siegman, A. W. 1961. Some personality variables associated with future time perspective. *Darshana* 1, No. 2:59–62.

Slater, P. E. 1970. *The pursuit of loneliness: American culture at the breaking point.* Boston: Beacon.

Sorokin, P. A. 1943. *Sociocultural causality, space, and time.* Durham: Duke University Press.

Sorokin, P. A., and Merton, R. K. 1937. Social time: a methodological and functional analysis. *American Journal of Sociology* 42:615–29.

Spiegel, L. A. 1958. Comments on the psychoanalytic psychology of adolescence. *The Psychoanalytic Study of the Child,* 13:296–308.

Spindler, G., and Spindler, L. 1965. Researching the perception of cultural alternatives: the Instrumental Activities Inventory. In *Context and meaning in cultural anthropology,* ed. M. E. Spiro, pp. 312–37. New York: Free Press.

Srole, L. 1956. Social integration and certain corrolaries: an exploratory study. *American Sociological Review* 21:709–16.

Stein, W. W. 1958. Andean Indian village. *Journal of Social Issues* 14, No. 4:5–16.

Stern, W. 1924. *Psychology of early childhood up to the sixth year of age.* Trans. by A. Barwell. London: Allen & Unwin.

———. 1938. *General psychology from the personalistic standpoint.* Trans. by H. D. Spoerl. New York: Macmillan.

Stone, L. J., and Church, J. 1957. *Childhood and adolescence: a psychology of the growing person.* New York: Random House.

Stouffer, S. A., Lumsdaine, A. A., Lumsdaine, M. H., Williams, R. M., Jr., Smith, M. B., Janis, I. L., Star, S. A., and Cottrell, L. S., Jr. 1949. *The American soldier: combat and its aftermath,* vol. II. Princeton: Princeton University Press.

Strodtbeck, F. L. 1958. Family interaction, values, and achievement. In *Talent and society: new perspectives on the identification of talent,* D. C. McClelland, A. L. Baldwin, U. Bronfenbrenner, and F. L. Strodtbeck, pp. 135–94. Princeton: Van Nostrand.

Terman, L. M., and Miles, C. C. 1936. *Sex and personality: studies in masculinity and femininity.* New York: McGraw-Hill.

Tomkins, S. S. 1962. *Affect—image—consciousness,* vol. I. New York: Springer.

Tresemer, D., and Pleck, J. 1972. Maintaining and changing sex-role boundaries in men (and women). Paper presented to conference on "Women: Resource for a Changing World." Cambridge: Radcliffe Institute, April 17–18.

Vernon, P. E. 1969. *Intelligence and cultural environment.* London: Methuen.

Vincent, J. W., and Tyler, L. E. 1965. A study of adolescent time perspectives. Proceedings of the 73rd annual convention of the American Psychological Association.

Wallace, M., and Rabin, A. I. 1960. Temporal experience. *Psychological Bulletin* 57:213-36.

Wax, M. 1962. The notions of nature, man, and time of a hunting people. *Southern Folklore Quarterly* 26:175-186.

Werner, H. 1948. *Comparative psychology of mental development.* New York: Science Editions.

White, R. W. 1960. Competence and the psychosexual stages of development. In *Nebraska symposium on motivation,* ed. M. Jones, pp. 97-141. Lincoln: University of Nebraska Press.

———. 1964. *The abnormal personality.* 3rd ed. New York: Ronald.

Whorf, B. L. 1956. *Language, thought, and reality,* ed. J. B. Carroll. Cambridge: M. I. T. Press.

Whyte, W. F. 1943. *Street-corner society: the social structure of an Italian slum.* Chicago: University of Chicago Press.

Winterbottom, M. 1958. The relation of need for achievement to learning experiences in independence and mastery. In *Motives in fantasy, action, and society,* ed. J. W. Atkinson, pp. 453-78. Princeton: Van Nostrand.

Wohlford, P. F. 1964. Determinants of extension of personal time. Ph.D. Dissertation, Duke University.

———. 1966. Extension of personal time, affective states, and expectation of personal death. *Journal of Personality and Social Psychology* 3:559-66.

———. 1967. Extension of personal time in TAT and story completion tests. *Journal of Projective Techniques and Personality Assessment* 32:267-80.

Wyatt, F. 1964. In quest of change: comments on Robert Jay Lifton's "Individual Patterns in Historical Change." *Comparative Studies in Society and History* 6:384-92.

Yerkes, R. M. 1943. *Chimpanzees: a laboratory colony.* New Haven: Yale University Press.

Zawadski, B., and Lazarsfeld, P. 1935. The psychological consequences of unemployment. *Journal of Social Psychology* 6:225-51.

Zentner, H. 1966. The social time-space relationship: a theoretical formulation. *Sociological Inquiry* 36:61-79.

Name Index

Stern, W., 32, 71
Stone, L. J., 72
Stouffer, S. A., 22
Strodtbeck, F. L., 105, 117, 190

Terman, L. M., 103
Thant, U, 203
Thompson, W. R., 7, 8
Tomkins, S. S., 13
Tremblay, M., 189–90
Tresemer, D., 104
Tyler, L. E., 97

Ullman, A. D., 7, 26

Vernon, P. E., 77, 90
Vincent, J. W., 97

Wallace, M., 86
Watts, J. W., 14–16
Wax, M., 166

Weaver, C. R., 128
Weaver, E., 127
Weaver, G., 128–30
Weaver, M. C., 129
Weaver, M. R. C., 127–28
Werner, H., 168
White, R. W., 18, 73
Whiting, J. W. M., 165
Whorf, B. L., 171, 172
Whyte, W. F., 192
Williams, G. W., 129, 130
Wineman, D., 19–20
Winterbottom, M., 117
Witty, P. A., 93
Wohlford, P. F., 11, 23, 106
Wyatt, F., 86

Yerkes, R. M., 6–7
Youtz, R., 206

Zavalloni, M., 213
Zawadski, B., 24
Zentner, H., 172

Subject Index

Modernization
  psychological consequences
    of, 197–217
  of traditional societies,
    176–83
Money Game, 113, 114, 118
Mortality, reducing infant, in
  traditional societies, 176–
  77

Non-linear time in traditional
  societies, 164–69
Nova Scotia poor, 189–90
Nuer (tribe), 166

Old, the, life study of, 36–68
Out Island Bahamas villagers,
  188–89

Past, the, see Time
Pawnee Indians, 166
Peasants, life planning by, 10
Peer-group activities, 89
Peruvian Indians, 181
Plans, as motivating force, 9, 10
Pleasure principle, 32
Poverty
  Colombian, 24
  culture of, 183–84, 186–87,
    195–96
  Ghanaian, 213, 214
  Nova Scotia, 189–90
  Puerto Rican, 183

Pre-marital intercourse, 194
Present, the, see Time
Prostitution, life study of,
  124–58
Psychiatric patients, levels of
  anxiety in, 29–30
Psychic mobility, 179
Psychotic patients, lobotomies
  on, 14–16
Psychopathic personality, 185
  anticipations and, 18–20
Puberty, vocational attitudes
  and onset of, 93
Public opinion surveys, mental
  ability and, 26
Puerto Rican poor, 183
Purposeful behavior, capacity
  for, 7

Reality principle
  defined, 32–33
  growth of, 92
  primary function of, 73
Reciprocity, ethic of, 191–93
Renaissance, time concepts of,
  174
Repetitive cycles, projection of,
  168–69
Representation of objects in
  determining behavior, 9
Role demands, 87–90
Roles, middle class dominated
  by, 185
Russia, students favoring
  showdown with (1951),
  26–27